Museums of Modern Science

Museums of Modern Science

SVANTE LINDQVIST, EDITOR

MARIKA HEDIN AND ULF LARSSON, ASSOCIATE EDITORS

Nobel Symposium 112

SCIENCE HISTORY PUBLICATIONS/USA

2000

First published in the United States of America
by Science History Publications/USA
a division of
Watson Publishing International
Post Office Box 493, Canton, MA 02021-0493

Library of Congress Cataloging-in Publication Data
Nobel Symposium (112th : 1999: Stockholm, Sweden)
Museums of modern science : Nobel Symposium 112 / Svante Lindqvist, editor;
Marika Hedin and Ulf Larsson, associate editors.
p. cm.
Includes bibliographical references and index.
ISBN 0-88135-299-3 (alk. paper)
1. Science museums—Congresses. 2. Communication in science—Congresses. 3. Museum techniques—Congresses. I. Lindqvist, Svante. II. Hedin, Marika. III. Larsson, Ulf, 1965– IV. Title.

Q105.A1 N63 1999
507.5—dc21 00-029680

Archives of the Nobel Museum 1
ISSN: 1404-7586

Light bulb photographs on title page from the exhibition catalog of *Edison: Lighting a Revolution, The Beginning of Electric Power* (September 1979). Additional information is available from the Electrical Collections, National Museum of American History, Smithsonian Institution.
Courtesy the Smithsonian Institution

Designed and manufactured in the USA.

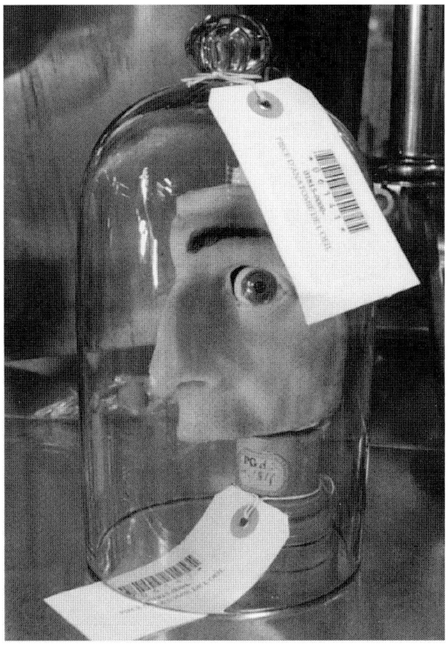

Contents

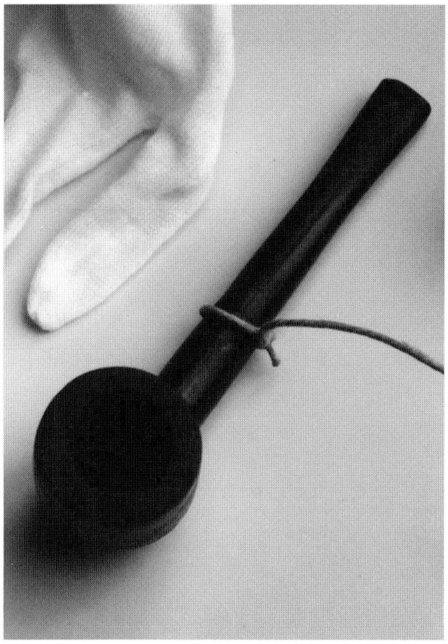

Introduction: Einstein's Pipe

To attempt to explain modern science is a matter of growing urgency. Many scientific issues need to be explained to, and seriously debated among, a larger public—to counter current trends of anti-science and non-rationality based on lack of knowledge or lack of interest (or both). More generally, the challenge is to attempt to explain and awaken an interest in matters of great complexity in a world where any serious message is out by peril of being drowned in the media cacophony.

The challenge is, bluntly put, to find an alternative to, on the one hand, the traditional museums of science with their instruments of brass in glass cabinets, and, on the other, the glossy hands-on exhibits presented out of context in many modern science centers. But even if this should succeed, the question remains how many are affected—intellectually and emotionally—and whether or not these attempts are studied only by those already converted. In order to reach new audiences, it is also important to place science in a broader cultural perspective: to present scientific development as an integral part of a more general cultural and social development, to show its many interfaces with society at large. As we may know, attempts to attract the interest of the audience by appealing to their interest in controversial issues are not unproblematic. Witness the "Enola Gay" controversy, or the storm of criticism the exhibition "Science in American Life" at the National Museum of American History faced from the US physical and chemical societies.[1]

In general, the problem is to find a level of explanation that is neither so superficial as to lose the complexity of the relationships, nor so complicated as to be incomprehensible to the audiences one wants to reach. It is also crucial to find ways of addressing the controversial issues of science that occupy the public interest without running the risk that the exhibitions will be seen as either toadying to popular opinions or as hostile to science, that is, of avoiding the risk of being dismissed as representing vested interests or as having an anti-science bias. At the same time, the intellectual excitement of the scientific process has to be conveyed and made aesthetically attractive in order to reach new groups, to reach the indifferent through parallels with art, music, and literature.

Two large international conferences dealing with these issues have been held during recent years at the Science Museum in London. The first was *Museums and the Public Understanding of Science* in 1992,[2] and the second *Here and Now: Contemporary Science and Technology in Museums and Science Centres* in 1996.[3] The first conference was arranged in co-operation with the British Committee for the

Public Understanding of Science (COPUS), and the second with support from the European Commission Directorate General XII. The British conferences were large gatherings (the latter had, for example, almost two hundred participants) in which broad surveys were given of on-going projects at science museums and science centers around the world.

In 1998, the Nobel Symposium Committee approved a proposal by Professor Tore Frängsmyr of Uppsala University and Professor Svante Lindqvist of the Nobel Museum to arrange a Nobel Symposium on how exhibitions of various kinds can help to explain modern science. The intention was to continue the discussion begun at the two British conferences by inviting a small group of international experts, and by structuring the symposium thematically around a few crucial issues. The symposium was held at the Royal Swedish Academy of Sciences in Stockholm on May 29–30, 1999, and arranged by the Nobel Museum and the Academy's Center for History of Science. Twenty-one invited speakers and session chairs from abroad and forty-three Swedish observers participated in the symposium.

The symposium provided an excellent opportunity for Swedish scientists, museum curators, and science journalists to listen to a number of excellent papers by distinguished foreign colleagues, and to meet and make contact with them. In particular, the symposium was of great interest to those of the participants who have been involved in forming the plans of the Nobel Foundation to create a Nobel Museum in Stockholm.

To many around the world, the name of Stockholm is synonymous with the Nobel Prize, with science and culture, but as yet there are no physical signs in the capital—much to the surprise of our foreign visitors—that every year this city is a center of world attention. Therefore the Nobel Foundation has initiated a project that will hopefully eventually result in the building of a new, permanent Nobel Museum in Stockholm.

The Nobel Museum is intended to give a physical form to the whole Nobel phenomenon—something which will give it an all-year-round permanence so that it does not appear only as the object of a media flurry for a few brief weeks each autumn and winter. It is a matter of displaying a cultural heritage, of defining and giving form to the cultural heritage that is represented by nearly a hundred years of the Nobel Prize system.

And indeed, this cultural heritage includes the undercurrent of criticism that makes itself heard every year. (In 1997, for example, it was the American journalist Daniel Greenberg of the *Washington Post* who discharged a volley to the effect that teamwork should be rewarded rather than individual efforts;[4] in 1999 it was the British medical journal *The Lancet* which devoted an editorial to arguing that Nobel Prizes for Medicine should be awarded for clincial work rather than for pure, biochemical research.[5]) In these recurrent complaints about those who ought to have got the prize but didn't, and vice versa, and about areas where there ought to be a Nobel Prize but isn't, and so on and so on, the wheel is reinvented every year. Even if this criticism is in many respects the same from year to year, it, too, should be made visible, given a continuity and put in its context. It, too, is a part of this invisible cultural heritage.

Another, more important, objective is to make use of all the knowledge of the main trends of scientific development that exist within the Nobel system, enabling the new Nobel Prizes every year to be placed in their context. One should not, to put it light-heartedly, be caught napping every October by the news that some old geezer in California has discovered something very, very tiny and the statement that this is very, very big. To us laymen the news is seldom immediately comprehensible, but a Nobel Museum could offer a context for the annual award of new Nobel Prizes. What more could one do for the public understanding of science?

All this may seem fine, but there is no point in denying that there are some problems here which perhaps should not be underestimated. Namely, that when the Nobel Museum was instituted in 1998 it was a museum which lacked (1) artifacts, (2) money, and (3) a building. It was truly what André Malraux called *un musée imaginaire*. But problems are meant to be overcome.

As far as artifacts are concerned, the collections of the new Nobel Museum were admittedly somewhat meagre when the museum was instituted. In fact the collections consisted of only one object: the original will of Alfred Nobel, dated Paris, November 27, 1895, an artifact to which we promptly

Einstein's pipe.

Photo: Thomas Asplund 1998

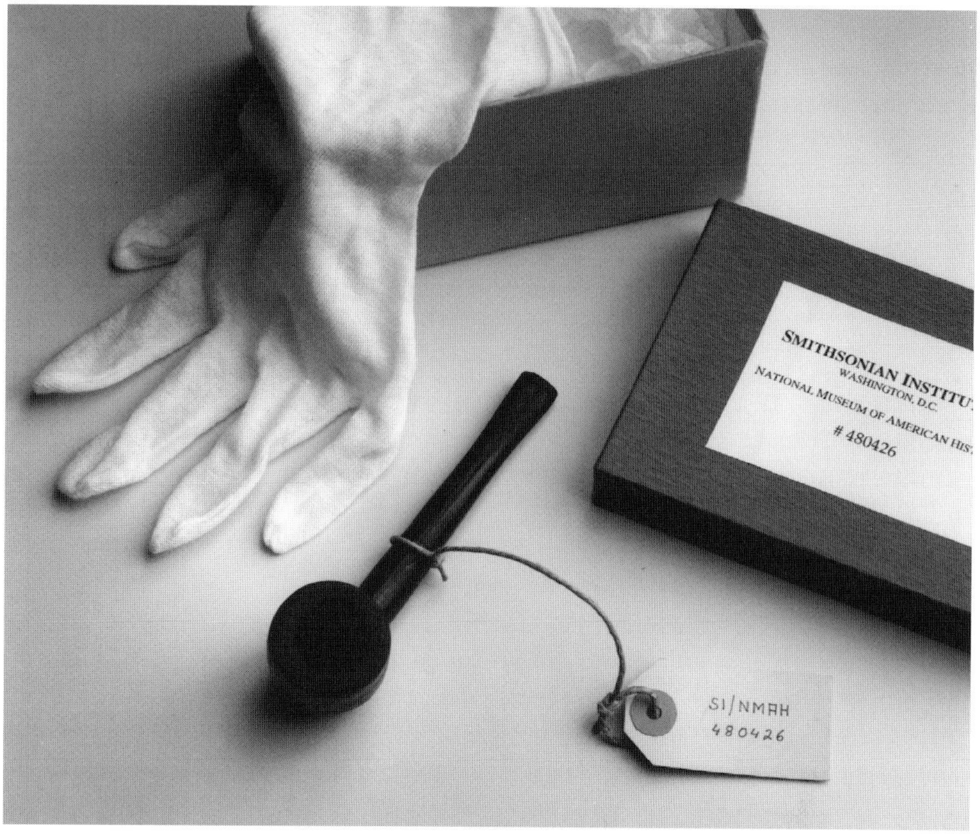

assigned Inventory Number 1. But we take heart from the fact that originally the collections of Sweden's most famous and frequently visited museum, the Vasa Museum, also consisted basically of just one object: the ship, this huge mass of dark oak which was the most advanced man-of-war, representing the cutting edge of contemporary technology, the most sophisticated example of early seventeenth-century naval engineering. At least it was for twenty minutes, the duration of its maiden voyage on August 10, 1628, before it sank like a stone. And if one of the greatest technological catastrophies in Swedish history can form the basis of our most successful museum and a constant source of national pride, surely the idea of a Nobel Museum—whose name represents a far more successful enterprise, one must admit—should have some potential?

Among the objects connected with American Nobel prizewinners that are in the collections of the Smithsonian Institution in Washington, D.C., is a pipe that belonged to Albert Einstein (*see illustration*). Now admittedly Einstein was a Swiss citizen when he was awarded the Nobel Prize for Physics in 1921, and it was not until 1940 that he obtained dual Swiss and American citizenship. But the prestige of the Nobel Prize is such that nations and universities collect Nobel Laureates much as churches collect pieces of the True Cross. The latter, if put together, would be sufficient to build a respectable ship. The former could populate a fair-sized town.

This object—Einstein's pipe—gives rise to many speculations. Its unique authenticity inspires an almost vertiginous feeling. Imagine—the bowl of this pipe was filled and grasped by the very hand that signed the famous letter to President Franklin D. Roosevelt on August 2, 1939 that led to the Manhattan project, and ultimately to Hiroshima and Nagasaki. This mouthpiece was sucked reflectively by the very lips that first pronounced the famous formula $E = mc^2$. As a typical male attribute, this pipe could also be used in an exhibition based on all the more recent research that has demonstrated Albert Einstein's complex and patriarchal relationship with his first wife and fellow-student, the relatively unknown Mileva Maríc. (Their first child, Liserl, born outside the marriage, was adopted immediately after birth, reputedly by a baby-farmer in whose care she soon died of smallpox.) But how can this artifact be used to explain the development of science? Or the interaction of science and culture? Or, more generally, the interaction of science and society in the twentieth century?

This prompts the question of how modern science might be explained and portrayed in the Nobel Museum. A Nobel Museum is of course, limited to twentieth-century developments. This museum will, by definition, be faced with the issue of how to explain "modern" science, since the first Nobel Prize was awarded in 1901. Furthermore, since new prizes at the forefront of science are, in a manner of speaking, "added to the collections of the museum" each year, they, too, need to be explained to a larger audience.

Here the traditional method of many science museums that depict the scientific revolution of the seventeenth century will simply not be adequate—that is, the practice of placing scientific instruments of brass and hardwood in glass cases and illuminating them against a background of green velvet. The LEP (large electron positron storage ring) at CERN near Geneva has a diameter of 27 kilometers. It would be very tricky indeed to get it into a glass case, and it would need an awesome quantity of green velvet.

To what extent should the Nobel Museum build up its own collections, and according to what principle? It is at any rate obvious that objects such as Einstein's pipe are not sufficient to demonstrate, for example, the large scale of today's "Big Science," the international networks of science, or the troubled relationship between research and those who finance it. No amount of pipes, no quantity of chips from the True Cross are likely to solve the problem.

Here the plans of the Nobel Museum for a larger system of itinerant exhibitions being shown abroad create a special problem, for if the number of identical exhibitions is more than one, the objects lose their claim to uniqueness. There can, I suppose, be only one original will of Alfred Nobel. But on the other hand, many of the icons of twentieth-century science that are on display at museums are mere facsimiles. For example, the model of the DNA molecule by Crick and Watson, "the double helix," at the Science Museum in London is only a reconstruction, manufactured long afterward by former colleagues at the laboratory in Cambridge of laboratory pieces, some of which were used by Crick and Watson.[6] And in the same way, the table with the experimental arrangement by which Otto Hahn and his colleagues are said to have discovered nuclear fission in 1938—which has a place of honor in the Deutsches Museum in Munich—is only a reconstruction.[7] All this becomes apparent only on reading the small print and, as an illusion, it works but then what is a unique artifact? This pipe, for example (*see illustration*), no longer has the same charm now that I admit that it is simply any old pipe, one that I bought in a second-hand shop in Stockholm. Or as René Magritte said "*Ceci n'est pas une pipe*." Theoretically, of course, it may have belonged to Einstein, but it is not very likely.

For many years, I had the privilege of being a pupil of Torsten Althin (1897–1982).[8] He had been Director of Sweden's *Tekniska Museet* for 38 years (1924–1962), and was now teaching the History of Technology at the Royal Institute of Technology, the engineering school in Stockholm. On solemn occasions, Althin always wore *Der goldene Ehrenring des Deutsches Museum*, a thick, heavy ring of solid gold adorned with the symbolic owl of the Deutsches Museum. He used to hum aria tunes from *Die Meistersinger*, and he often spoke of how he once saw Richard Strauss conduct during a concert at the opening of the Deutsches Museum in 1924. In the 1970s when I met him, Torsten Althin was in his late seventies and somewhat of a legend in the Swedish museum world. Ten or fifteen years earlier, in the early 1960s, he had been at the peak of his reputation in the international museum world, but he had by now outlived most of his many friends in ICOM and the museums of post-war Europe and the United States. He drew a constant stream of visitors to our little office, with whom he enthusiastically shared his encyclopedic knowledge and many anecdotes of the past. He always used to tell them that "one single real object tells far more than a thousand words."[9] This, however, may have been true of the industrial museums of the early twentieth century, but I would like to argue that this—the uniqueness of Einstein's pipe—is not the issue that faces museums of modern science, and the future Nobel Museum in particular. The issues are rather those about which Roald Hoffmann, who was awarded the Nobel

Einstein smoking the same.

Transglobe/David Eisendraht, Jr.

Prize in chemistry in 1981 and who has served on The Smithsonian Institution Council, has recently written:

> The Nobel Prizes recognize artistic and scientific ideas, knowledge and understanding. Artifacts too—accelerators, therapeutic techniques, molecules that heal or may hurt, books and poems. But the ideas more than the objects are what the Nobel Institution celebrates.
>
> Most museums preserve and interpret objects. A Nobel Museum's mission is unique: its "acquisitions" are just those ideas that have served human beings and given them understanding and spiritual, yes, spiritual satisfaction. It is not at all evident how one might transmit to the future the intangible, the stuff of the mind. The way of preserving and passing on ideas—stronger than any cannon, more precious than any royal treasure or painting—the way must be inherently different from that of objects. I think it must be through the widest dissemination of ideas, across cultures, in every possible medium.[10]

The organizers of the Nobel Symposium no. 112, "Museums of Modern Science," would like to express their gratitude to the Nobel Symposium Committee for the opportunity to arrange this symposium. We are also grateful to the Royal Swedish Academy of Sciences for their generous hospitality, and this includes, among others, Professor Erling Norrby, Secretary General of the Academy; Mrs. Marianne Nilsson, Chief Accountant; Mrs. Karin Hedberg, Head Housekeeper; and Mr. Ulf Linderoth, Supervisor. We would also like to express our most sincere thanks to Mr. Neale Watson, President and CEO of Watson Publishing International, and his staff for their wholehearted enthusiasm during the production of this volume, and for its elegant design. And finally we would, on behalf of all the Swedish observers who had the benefit of attending this Nobel Symposium, like to thank the speakers—the authors of this book—who so generously shared with us their knowledge of and experience with museums of modern science, to which this volume witnesses.

Svante Lindqvist
February 29, 2000

NOTES

1. Arthur Molella and Carlene Stephens, "Science and Its Stakeholders: The Making of 'Science in American Life,'" in: *Exploring Science in Museums*, ed. Susan Pearce, New Research in Museum Studies no. 6 (London: Athlone, 1996), 95–106; Robert C. Post and Arthur P. Molella, "The Call of Stories at the Smithsonian Institution: History of Technology and Science in Crisis," in: *Icon. The Journal of the International Committee for the History of Technology* (1997) 44–82; William S. Pretzer, ed., Special Section: "The Last Act: Reviewing Public History in Light of the *Enola Gay*," *Technology and Culture* 39 (1998), 457–488.
2. John Durant, ed., *Museums and the Public Understanding of Science* (London: Science Museum, 1992).
3. Graham Farmelo and Janet Carding, eds., *Here and Now: Contemporary Science and Technology in Museums and Science Centres* (London: Science Museum, 1997).

4. Daniel Greenberg, "The Annual Overrating of the Nobel Prizes," *The Washington Post* Sept. 24 , 1997.
5. *The Lancet*, no. 9188, vol. 354, Oct. 23, 1999.
6. Ann Newmark of the Science Museum, London, visited Herman Watson in Bristol in 1976, where she discovered he had the Watson and Crick metal plates. He agreed to let the plates come to the Science Museum in return for some replacement plates which were made for him in the Museum workshops. Farooq Hussain, then at Kings College, London, agreed to reconstruct the Watson and Crick model, using these original metal plates with standard model building pieces. He started in late 1976 and completed the model in time for it to be installed in the Chemistry galleries which were being refurbished at that time. The Chemistry galleries opened in March 1977 with the DNA on display for the first time. (I am indebted to Robert Bud and Ann Newmark for this information.)
7. Around 1962, when Klaus Biedermann was a graduate student and assistant at the Technical University of Munich, working at a newly founded department outfitted with recycled furniture, two gentlemen from the Deutsches Museum visited him. They had a carpenter's ruler with them and a photo and were searching for an old table, suitable as a replica for Hahn's and Strassmann's experimental setup. He could help them. (I am indebted to Professor Klaus Biedermann of the Royal Institute of Technology, Stockholm, for this information.
8. Svante Lindqvist, "An Olympic Stadium of Technology: Deutsches Museum and Sweden's Tekniska Museet," *History and Technology* 10 (1993), 37–54; *idem*, "Des olympiades de la technologie: Le *Deutsches Museum* et le *Tekniska Museet*," in: Brigitte Schroeder-Gudehus, ed., *La Société industrielle et ses musées: Demande sociale et choix politiques 1890–1990* (Paris: édition des archives contemporaines, 1992), 119–149.
9. Torsten Althin, "An Intellectual Autobiography," *Technology and Culture* 20 (1979), 589.
10. Roald Hoffmann, "A Mission Statement for the Nobel Museum," *Newsletter of the Nobel Museum* No. 4 (Autumn), 1999, 2–3.

The Issues: How to Explain Modern Science and How to Reach the Indifferent

Museums in the New Millennium

Neil Cossons

As we approach the end of the century and I lapse into my anecdotage, it seems I should begin with an anecdote. The story was told originally by that eminent historian of technology, the late Charles Singer (1876–1960). He is perhaps best known as the principal editor of the great five-volume *Oxford History of Technology*, published in the 1950s and '60s. In his Dickinson Memorial Lecture,[1] given at the Science Museum, London in 1954, Singer recalled the Second International Congress on the History of Science, held at the museum in 1931.[2] The Science Museum was then of course in new premises, for although the museum itself dates from 1857, it had occupied the present building on Exhibition Road only some three years previously.[3] At that time the museum was firmly in the capable hands of my distinguished predecessor, Sir Henry Lyons (1864–1944), who had been Director since 1920.[4]

In his address, Singer referred to the issue of hospitality at the congress:

> In those days Governments considered it no part of their duty to foster learning and still less to maintain the prestige of British learning. To offer to foreign visitors a cup of tea, paid for by Government funds, was considered slightly disreputable and likely to lead to the slippery slope of corruption. However, Sir Henry bravely risked the slope. Nothing in particular happened . . . Sir Henry placed the Museum premises at our disposal. Our foreign guests combined with us to consume hogsheads of Treasury tea.[5] The Empire did not collapse and nobody accused our amiable host of corruption.

Singer then spoke of the contribution made by the tea toward the success of the congress and went on:

> It is not true that these remarkable results were achieved entirely by tea. Buns played their part. If my memory serves, Sir Henry told us that he was willing to risk tea but that his position was not secure enough to provide Government buns. I think the buns for the foreigners were paid for by the English members of the Congress. Perhaps a careful research into the accounts of the Congress may reveal the source of the bun fund. But whatever the origin I am sure that the bun fund was a good national investment.

I use Singer's anecdote as a demonstration that, despite what some Jeremiahs may say, things have changed, overwhelmingly for the better, in museums and the field of scholarship in the seventy years or so since that congress in South Kensington. My belief is that they will continue to do so.

But on the eve of the new millennium few people might identify advancements in the quality of cuisine as one of the great indicators of change. In several senses they would be wrong. Increases in net disposable income, and time in which to spend it, have been among the decisive characteristics that exemplify modern western democracies and set them apart from their pre-industrial forebears and less developed contemporaries. Both net time and net income have grown consistently in the last half-century, allowing the many to enjoy privileges and opportunities until recently restricted to the few. Free time can now be enjoyed as never before, on eating out, on travel and holidays, on sport, and of course, on cultural pursuits, of which museums are an increasingly important part.

I make this point now because it is perhaps too easy for us, when contemplating the future of museums, to subscribe to simplistic clichés about change. Many in museums find change disturbing if not actually threatening, especially when that change is imposed from outside. This is sometimes the case when we consider the impact of science and technology and, in particular, information technology. The force of scientific and technological change is undoubted, but by far the most significant factors determining the nature of tomorrow's museums will be cultural, social and economic, not necessarily in that order. Issues of time—and competition for time—increases in mobility and the improvements in access that this affords, increased choice in the deployment of disposable income; all will of course exert profound influences on how the museums of tomorrow develop and progress. But, above all, the main determinant of change will be, as now, the interests, aspirations and predilections of the cultural and political elites who govern, direct and run museums. For museums, perhaps more than any other institutions of learning, of scholarship, of enlightenment or inspiration, are the product of the providers. The people who use them, at least for the time being, are bit players in the bigger performance. That will change as the new millennium dawns. Tomorrow's successful museums will be those that have a clear understanding of the values that they represent, a vision of where they are going and an ability to communicate that vision to the wider world. They will also have the capability of managing change, creatively and strategically. Most importantly, they will engage their audiences in determining their policies and measuring their performance.

What I should like to review in this paper are some of the recent trends in museums, both globally and from our own experiences in the United Kingdom, and then to see how these fit with our perceptions of and predictions for the new century. It is peculiarly germane that we consider the future of museums here and now, not only because the new millennium is almost upon us, and this demands that we pause and take stock, but, more importantly, because museums themselves are at a critical stage in their evolution. It is a problem of success. For the first time in the history of museums, we are seeing an upsurge of widespread although unfocused debate about what it is that museums think they are for. As yet that concern tends to be naïve and is often pretentious, but as it gathers momentum it is spilling out into the wider public realm. Positions are being staked out, not in the literature of the professionals—that is noticeably quiet on the matter—but right across the cultural and political media. Museums have at last arrived. They are seen to have a voice—attitude even—and so they have become the matter of legitimate public concern. The monopoly of the providers is for the first time being challenged. That must be a good thing. This is the price museums pay for the success and greater public visibility they have always wanted. Just as society has become more pluralistic, so has the provision of museums. But, most importantly, museums are perhaps for the first time being seen as the intellectual property of the people who use them and those people are increasingly wanting their likes and dislikes to be heard.

In the last twenty years, museums have moved from the twilight to the spotlight of public attention to assume a central position in the cultural firmament. The world wants museums. Throughout the developed—and increasingly in the developing—world, everybody believes they need to have museums. Given that the notion of the public museum is at least 250 years old, this is astonishing. Museums are an important and extraordinarily enduring part of our culture and our civilization. And the central purpose of a museum, in its underlying essentials, has hardly changed. Museums hold collections and reveal them to audiences. They are about objects and for people as they have always been. This mandate, to hold collections in perpetuity, is widely accepted and understood. In fact, the public's expectation that museums care properly and professionally for the stuff we leave with them is stronger than many of those who run museums generally realize. If Aunt Agatha leaves her favorite teapot to the local museum, it is done in the firm belief that it will be there for her daughter and granddaughter to see and enjoy. And, of course, it will always be on public display. Collections define the museum's right to permanence. It is a potent and persuasive argument.

In fact, such is the power of the museum idea that almost any place to which people come to see and understand aspires to be one. In the United States between $4 and $5 billion has been spent on new museums in the last decade.[6] Some 600 new art museums have been built since 1970[7] and in other fields the total is even higher, from the Holocaust Museum in Washington to a shrine in Clinton, Oklahoma "for the artifacts of Route 66, a nostalgia-driven trip down the Depression-era cross-country highway of deco diners and mom-and-pop motels."[8] And today virtually every major city in the United States has a science center, some 300 in all, attracting 115 million visitors a year and with 700,000 families subscribing to their membership programs.[9] As the term *museum* has lost its pejorative connotations, so it has widened its meaning and acceptability. Today it can embrace everything and anything, from collector's *cabinet* to hands-on science, from *wunderkammer* to waxwork. In fact, the museum has become such an open-ended compendium of everything we may wish to have around us that definition has become a real issue.[10] And, as ultimate sanctification, the word *museum* has become adjectival—"museum value" and "museum quality" have joined the enduring "museum piece" in the lexicon of meanings.

In Britain the trend is the same.[11] The popularly quoted statistic—that a new museum opens, on average, every two weeks—seems to have held good since the 1970s. Although there have been closures, these are few in number and, as yet, none has affected any other than the smallest of museums. But the number of visitors to museums, while continuing to grow, is not expanding at the rate that museums are opening. As a result, between 1978 and 1988, the *average* number of visitors to museums in Britain fell from 72,000 per annum to 48,000.[12] These figures of course hide other trends, most notably the significant increases in the numbers visiting large museums, especially in London. Broadly speaking, the large museums appear to be doing well, as do the very small and most, but not all, of the new. The museums that are in difficulty are the middle-sized and middle-aged, short of capital for renewal and of a sub-optimal scale of operation. While the smallest will always survive, on a mixture of ingenuity and adrenalin, and the large continue to re-invest in their infrastructures and the quality of their shows, the mid-sized flounder.

The impetus for much of this new museum mania derives from two concurrent and worldwide infatuations—for museums of contemporary art and for museums of inter-active science. I should like to spend a little time looking at this peculiar phenomenon because it goes to the heart of many of our cultural attitudes toward and perceptions of what museums are for and, more importantly, whom they are for. Many of the metaphors about change and whether that change is for better or for worse can be traced through the rhetoric of the contemporary debate and offer a little insight into the way museums will evolve in the twenty-first century.

In the case of the art museum, in the last decade or two we have seen it assume a crucial iconic position in civilized urban culture that far transcends its role simply as a place to present art. It is in the art museum that the relationship between architecture and art is peculiarly sensitive and symbiotic, further complicated by the fact that two major arts are involved, one housing the other so to speak, and are sometimes in conflict. We see James Stirling's Staatsgalerie in Stuttgart; here in Stockholm, Raphael Moneo's Modern Museum; Richard Meier's new Getty and his other works in Barcelona and Frankfurt; Tadao Ando, with a number of art museums to his credit in his native Japan, is building in Fort Worth, Texas; the Los Angeles Museum of Contemporary Art is the work of another Japanese, Arata Isozaki, and the Swiss architect, Mario Botta, was selected for the San Francisco Museum of Modern Art. The list is endless. Most significant of all, of course, has been Frank Gehry's Guggenheim in Bilbao. These signature buildings by signature architects, housing in many cases substance that is altogether evanescent, have become in their own right the most significant of cultural objects. They are museums judged as much by their architecture as their art. As the German critic Claus Kapplinger has pointed out, it is the building rather than the collection—even supposing that there is a collection—which has become the main attraction.[13]

And if the building is the exhibit, the name is the brand. So, Guggenheim takes its place with Gucci or Armani, or for that matter with Coca-Cola, as an international label. For $320 million the Basque government has bought into the Guggenheim brand, gambling the future of the city on culture and tourism. It seems to be paying off. In one step Bilbao has emerged from seedy post-industrial obscurity onto the world stage. But Bilbao does not own the brand. Who does, who protects its reputation, and who franchises it to others is another matter. But for the time being cities around the world with little going for them crave the name Guggenheim to lift them from the shadows. It is impossible to imagine that this will not continue to be one of the great museum trends of the new century. It may have a little to do with culture. But it has much to do with power and position.[14]

These museums offer new kinds of spaces with which, irrespective of what hangs on their walls or stands on their floors, people wish to be associated, as visitors or politicians, the aspiring young, corporate patrons, or the wannabes of tomorrow. It is something that those of us interested in science museums, and science and technology education, might observe with some care. There are things that we might wish to adopt or adapt from this peculiarly omnipotent cultural occurrence, about language, tone of voice, about hierarchies, cultural positioning and social acceptability. For many, however, it is the whiff of elitism and the fact that these new art museums, despite what their protagonists might say, are not for the deprived, the underprivileged, or the uneducated that sets them apart. A gallery director friend of mine prohibits use of the words "exciting" or "fun" in any literature or publicity material about his gallery. He believes—understandably and almost certainly rightly—that their use would demean the aspirational values he has so carefully nurtured. Sir Roy Strong, former Director of the Victoria & Albert Museum, put it slightly differently but no less explicitly when he talked of his museum as a place where people could "sip martinis among the Bellinis." And for those who are not among the cognoscenti, museums of contemporary art can pose severe issues of understanding. Tourist boards find them peculiarly difficult. The best *Stockholm.com*, for example, can manage for the Moderna Museet is: "Very trendy museum. Very trendy web-site."[15]

Clearly the museum culture comes in a variety of flavors. Not all of these are acceptable to everybody. As a result, the voices of dissent have been getting noisier both from within the museums community and from the outside world. As museums have become increasingly influential, so the intensity of the debate about them has grown. So far, more heat than light has been generated. The battle has been raging since the mid-1980s. There are few signs that peace is about to break out. And, although Britain and the United States have provided the main battlegrounds, evidence of collateral damage is widespread and not only in the English-speaking world. At first sight the debate is about the defense of scholarship in the face of populism. But the divide is much deeper than this and the issues more complex. These culture wars are in essence between those who believe they should have unchallenged authority over the nature of the museum experience and others who are prepared to share the notion with a wider public. Generosity of spirit is not much in evidence; the growing literature offers a rich mix of fire and brimstone. Caroline Reinhardt, writing in *The Spectator*, is one of a succession of commentators who have entered the fray:

> There is something happening behind the scenes at the museum. A revolution has taken place in its philosophy, which would like to see the glass cases smashed. Today's museum aims to be genuinely populist. It welcomes—indeed seeks out—all sectors of the community, and eschews anything that smacks of elitism. Explanatory material (preferably using state-of-the-art technology) is pitched at the simplest possible level. And, above all, the new museum seeks to pull its head out of the historical sand to address issues in the contemporary world.[16]

Reinhardt drives home her point by quoting the Director of Tyne and Wear Museums, who believes that museums should play "a proactive role in con-

FIGURE 1. The Spiral. Daniel Libeskind's vivid new addition proposed for the Victoria & Albert Museum, London. This computer-generated image shows the view looking south along Exhibition Road.
Courtesy Victoria & Albert Museum

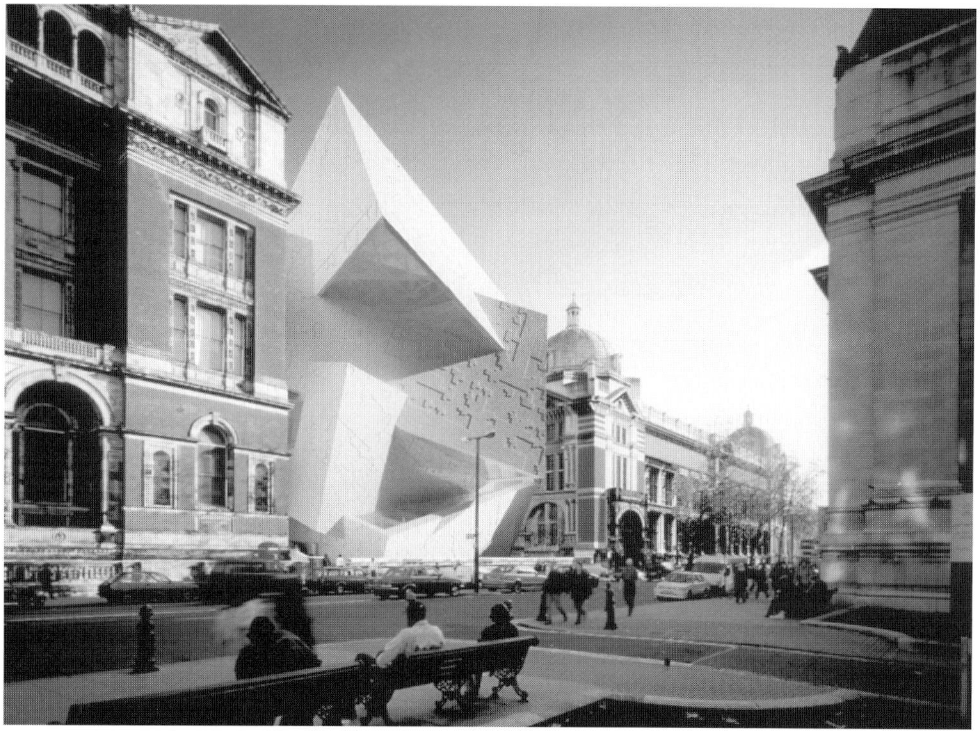

temporary societal issues . . . [and] act as an agent of social change."[17] Social historians, new technology, widening access, relevance to minorities, the "new museology," all share the burden of blame for making museums easy entertainment instead of offering intellectual challenge and demanding some effort on the part of the visitor. In the view of Heather MacDonald, the cancer at the heart of museums emanates from "the worst elements of America's academic culture" ". . . smirking irony, cultural relativism, celebration of putative victims, [and] facile attacks on science." Prime suspects are the curators in the Smithsonian and their "embrace of postmodern theory and identity politics."[18]

In 1990 the Royal Society of Arts was persuaded to stage a show trial of the offenders, mainly in their absence. The press notice was revealing, identifying a wide set of anxieties.[19] It asked whether scholarship was being sacrificed to the demands of storage, conservation and popularization. Is the ". . . controversy . . ." ". . . a necessary consequence of government under-funding or the result of a quite separate thrust towards treating the command structure of a museum as though it were identical to that of a business? Should the scholar-curator still play the central role in a museum or should he be no more important than the full-time manager, the conservationist (sic), the accounts and marketing man?" Here managerialism and commercialization were added to the list of more traditionally accepted perpetrators of crimes against culture, in part fueled by the uniquely British debate about whether or not museums should charge their visitors at the point of entry.[20] The association in the minds of many, of managerial tendencies, charging for admission and populism, working in conjunction to threaten scholarship and traditional curatorial values, formed a repetitive liturgy throughout the conference. What nobody was prepared to say in anything other than carefully encrypted code was that these people were declaring their ownership of museums and asserting a right to public money in support of that claim. And, as guardians of museum culture, they dispensed museum values to their audiences on terms that only they could determine.

If this is elitism, then it has its advocates, although in espousing publicly such unfashionable doctrines other flaws become unwittingly exposed. Interviewed in *The New Yorker*, Philippe de Montebello, longstanding and outstanding Director of the Metropolitan Museum of Art in New York, makes a spirited defense of the special values and qualities of a museum:

> Too many museums today are trying to become like theme parks or upscale shopping malls . . . There is a complicity in the notion that you have to compete with Disney and this is *tragic.* If you start to compete on the level of the theme parks, you are going to lose, because they can do it so much better.[21]

Few would argue with that, but he then goes on to describe the museum's audience, as he perceives it:

> . . . I hate to call it a mass audience. The numbers are not high enough, frankly, at five million visitors. If you begin to decorticate those numbers, you find that there are a great many repeat visitors—I don't mean people who come once or twice but people who come six or seven times, or even twenty times, a year. I would be surprised if our audience is as high as one million *different* human beings. And it's a more sophisticated audience than you get at a museum like the Louvre or the British Museum. A lot of the people that you see popping in and out of the Louvre have been dumped there with no foreknowledge. . . .

The Met's audience, in de Montebello's view, does not need to be sold on the art, because the audience is, *a priori*, a cultural elite.

There is complicity here too. Museums are value-driven rather than bottom-line-driven organizations. They represent quality but, when the chips are down, they are more than happy to see their achievements measured in quantity. Visitor figures are the common currency of this debate. They can be used to present a picture of widespread, but wholly spurious, community participation. This can be highly persuasive when claims are being made on the public purse, and thus the taxpayer becomes an unwitting accomplice to the myth that big numbers mean wide use by the population at large. In fact, the public is often supporting a narrow fellowship of the well heeled, whom they are subsidizing to enjoy frequent and regular use of their own domain.

The perceived values of the science center or science museum, on the other hand, are entirely and almost perversely different. To their detractors they are seen as little more than fun palaces. Here children with short attention spans dart noisily from exhibit to exhibit in uncontrolled Brownian motion. Others see issues of democracy at stake and view science centers as offering an essentially participative and accessible environment clearly focused on the needs of their audiences. Engagement with and empowerment of the young visitor, rather than didactic teaching, and strong links into the community are often the prevailing characteristics. Janet Daley, writing in *The Independent* at the height of the furore in Britain over threats to the traditional values of museums, set out the nature of the dilemma with characteristic pithiness:[22]

> . . . the director of the Victoria and Albert Museum . . . argued forcibly for a more accessible image for her museum. The "ace caff with quite a nice museum attached" advertising campaign and her exhibitions for non-specialists had brought a deluge of criticism. Shrieks of "crass" and "down-market" reverberated around South Kensington.[23] Walking past the Science Museum the other day, I noticed its latest advertising poster, which shows a lavatory cistern tank with an old-fashioned pull chain. The text above the picture reads "In 1939 Ladislao Biro designed the ballpoint pen. In 1908 William Hoover put his name to the vacuum cleaner, and in 1900 . . ." And there emblazoned on the red cistern are the words: "Crapper's Valveless Waste Preventer". Can one envisage the Tate advertising its Constables with the slogan; "If you thought hay was for rolling in . . . ?" Why should the V & A invoke such flak when it strains the bounds of solemnity and purism, while its neighbours . . . can be facetious, even prurient and—most significantly—openly democratic? The reasons seem to me to go to the heart of British attitudes towards science and the arts. When Mrs Esteve-Coll proposes contextual settings for objects and perhaps even an active element in exhibitions, she is berated for turning the museum into a historical theme park.[24] But walking into the newly redesigned Science Museum, I was engulfed in a sea of hands-on exhibits bleeping and blooping at me like so many Space Invader machines. No one apparently is offended at the idea that a science museum should be didactic because, I would suggest, there is no shame attached in British society to knowing nothing about science. . . . Science . . . is for schoolchildren, something one grows out of with maturity and the coming of civilised tastes. C.P. Snow had it wrong. It is not so much that the arts and the sciences are two cultures as that science is no part of culture at all and so this country remains chronically at odds with the twentieth century.

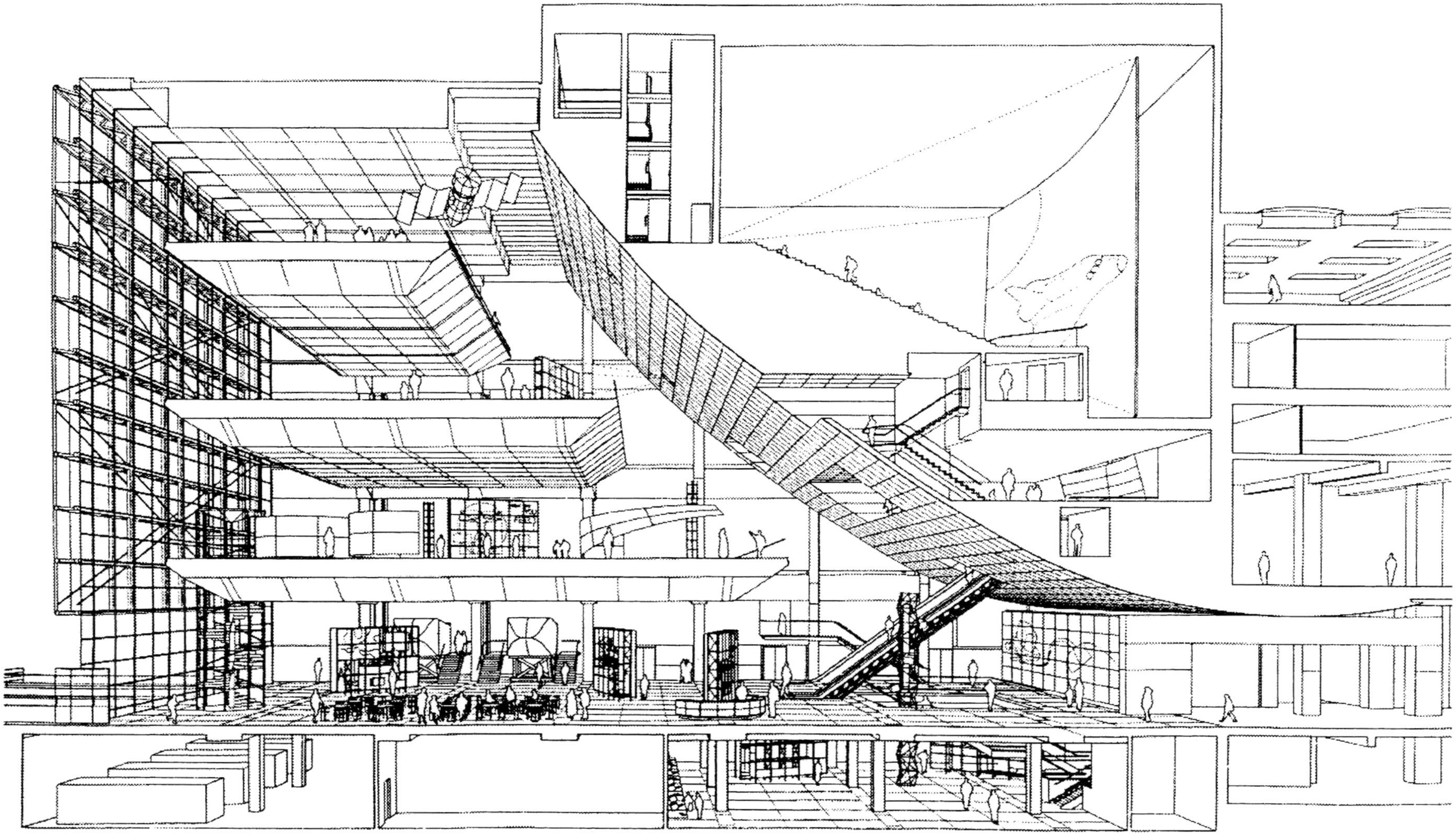

FIGURE 2. A Theatre of Science.

Schematic section through the new Wellcome Wing at the Science Museum, London. Opening in June 2000 the Wing is devoted to contemporary science and technology and is designed for frequently changing themes and exhibitions. Three exhibition floors and a 450-seat IMAX theatre float in a dramatic blue volume.

Architect: MacCormac Jamieson Prichard

Courtesy Science Museum

This view is not, I believe, peculiar to Britain. On the contrary, throughout the developed world the science museum—but more specifically the science center—has espoused a populist approach. In doing so it has acquired a downmarket image, highlighted in many cases by tired and worn out fittings, exhibits that should work but don't and a generally down-at-heel appearance of buildings and contents, all of which contrast markedly with the expensive finishes and high standards of maintenance of other museums. Poor visual design and literature that presents a well-meaning but essentially repetitive rhetoric aimed at appealing to children, educationalists and parents reinforces this still further. Janet Daley may well be right. When young people themselves view science as something they finished with as children, small wonder that puberty appears to be the great enemy of the public understanding of science. Science centers, set up to inspire and engage, may in fact be laying the ground for a conscious and forthright rejection of science by the young once they become aware of more appealing alternatives. Conscious targeting of adult audiences and, in particular, affording continuity across the critical age range of the late teens and early twenties would appear to be critically important.

In an increasingly visual world where—like it or not—style, design, distinctiveness, choice and quality are seen increasingly to matter, the determination of many science centers to face the other way seems like a blueprint for disaster or at least for cultural marginalization. Indeed, a high proportion of science centers may be condemned to suffer perpetual penury as a result of their failure to meet the rising expectations of their visitors. The then Head of Research and Development at *newMetropolis* in Amsterdam, James Bradburne, has gone so far as to argue, in the event rather prophetically, that science centers are by their nature inherently unsustainable.[25] Many of them will blame this on lack of funds, but the real culprit is lack of ideas, imagination or understanding of the needs and wishes of people. Money will always follow inspiration.

What does all this mean for museums of science, technology and industry—for what are usually referred to generically as science museums? First, there is the issue of collections. Around the world large multi-disciplinary museums of science and technology are relatively uncommon. The genre seems to have been a product of the industrial age and any nation with aspirations to world power status needed to have one. In Austria, Britain and Germany, for example, national museums of science and technology had all been established by the 1920s. The United States, anxious not to be left out, was thwarted in its own ambitious plans by the Wall Street Crash of 1929, but other museums of comparable stature were set up, in Dearborn, Michigan (by Henry Ford), and Chicago.[26] It is doubtful if many further examples will now be established and those that already exist show marked signs of ambivalence toward their historical holdings. What has caused them to pause and question their assumptions about the public worth of their collections has been the worldwide spread of the inter-active science center. The educational philosophies of the noted physicist and educator Dr. Frank Oppenheimer (1912–1985) pioneered in the Exploratorium in San Francisco,[27] and the stylish and immensely popular Ontario Science Centre in Toronto, opened in 1969, have provided the inspiration for innumerable such places around the world.[28]

These new science centers have eschewed historical collections in favor of apparatus with which visitors—and especially young visitors—can physically engage in order to demonstrate for themselves principles of science and technology. The idea is founded on the premise that first-hand experience with scientific phenomena will captivate ordinary people and stimulate original thinking about science. Although in a developed form the philosophy dates from the 1960s, and the decade of feverish educational reform in the United States after the launch of Sputnik in 1957, inter-active exhibits as such are much older. Examples could be found in the South Kensington Museum in the nineteenth century and the Palais de la Découverte in Paris and the Chicago Museum of Science and Industry in the 1930s. From the outset they were popular. The Children's Gallery in the Science Museum, London, opened in December 1931, was an important contributor to raising visitor numbers above one and a quarter million a year, a figure which then exceeded that of the British Museum and was second in Europe only to the Louvre.[29] Nevertheless, from the outset these places also attracted the opprobrium of the museum community. The *Museums Journal*, commenting on the Children's Gallery, said: "We could not help fearing that all this may be going too far and not quite in the right direction. . . . [The Science Museum's] ordinary exhibits already appeal so forcibly to the young that a trifle more elementary exposition is all that can be needed."[30] But the public was enthusiastic, as was the press. The *Daily Telegraph* noted that the Children's Gallery, far from having "degenerated into a bear-garden, as some Jeremiahs predicted, it has turned out one of the complete successes in making self-education thrilling."[31]

In the face of the inter-active revolution, the position of collections in those large science museums that have them has become unclear, both to the museums themselves and to their audiences. As a result, many big science and technology museums have betrayed their collections by allowing them to become marginalized in the minds of the public. Although they may have paid lip service to the value of their collections as a resource for scholarship, many museums have had neither the imagination nor narrative skills to bring them alive in the minds of the visitor. Galleries of murky, incomprehensible machines mean little to any but those who made or used them unless presented with style, understanding and vivid narrative. As the generations change so the problem worsens.[32] This is exacerbated by the inability, endemic in large museums, to renew so-called permanent exhibitions at anything like their rate of decay might demand.

Robert Bud's contention that "meaning is inherent in objects, and a museum is therefore a storehouse not just of objects but also of meanings" raises important questions of how far this is fully understood by museums themselves.[33] Why do museums do such a bad job of revealing those meanings to any but true believers? Is the museum a place for collections or a vehicle for ideas? Are the two compatible? The dilemma reflects a confusion about whether the museum is object-driven or issue-driven. Is it capable of using its collections to decode messages for its public or do its stories derive from issues of relevance to its various audiences and for which its collections may—or may not—offer some degree of illumination. For museums of science and technology, and especially for those dealing with contemporary issues, understanding scientific principle or process adds a further dimension to the complex equation of revelation. But new generations of curators and interpreters are working together to remedy these deficiencies. In Birmingham, for example, Discovery Centre at Millennium Point will place the outstandingly important collections of the, now closed, Birmingham Museum of Science and Industry in the context of the reviving fortunes of the city's central area and an audience for whom the traditional metal-bashing industries on which the city's rise and worldwide fortune was based are now little more than an historical footnote.

All this might lead us to think that museums have lost their way. Nothing could be further from the truth. For the first time, museums are questioning their own motives and, equally, they are being challenged and in some cases made accountable to the public and the agencies that fund them. This is perhaps one of the most invigorating of the changes in museum culture that we can look forward to in the new millennium. The problem, if there is one, is that some of those who work in museums don't like the answers they are hearing. As the zephyrs of new thinking strengthen to become if not a wind at least a stiff breeze, and start to blow away the cobwebs from some of the dingier corners of the museum world, grumpy fundamentalists in their protected habitats become exposed for what they are. Accountability threatens their autonomy.

Let me conclude by summarizing some of the changes in museums that we can expect in the new century. As museums are seen increasingly to matter, they will become the focus of more critical and analytical public debate, about their policies and their messages. Painful as this will seem to some, this scrutiny must be beneficial. In particular, museums of science and contemporary history will find themselves drawn into controversy as exhibitions penetrate the traditional confines of public comfort, resulting in challenges to their long-assumed and unquestioned authorities and freedoms. In this respect, the impact of the *Enola Gay* debacle goes far beyond the walls of the Smithsonian to affect museums worldwide.[34] But none of this will deter museums from tackling sensitive issues; on the contrary, they will increasingly push boundaries to the limits.

Meanwhile, public expectations of museums will continue to rise, as will intolerance of those that fail in their fiduciary responsibilities, to care for their collections and to make them accessible. Already, in Britain, the funding of national museums is tied to a basket of performance indicators agreed between the respective Boards of Trustees and the Secretary of State for Culture, Media and Sport.[35] These go well beyond crude visitor numbers to include, for example, measures of visitor satisfaction, percentages of collections held in environmentally secure storage and numbers of scholarly publications. And, just as sponsors over the last twenty years have placed on museums ever tougher demands for visibility and a return on their investment, so taxpayers will expect new standards of service and initiatives to widen access to communities currently denied the benefits that museums can offer. The critical questions are, can museums serve new audiences without aggravating the old ones and will greater openness and accessibility threaten public perceptions of their authority and thus public willingness to fund them.

New popularity, challenging messages and more intense public scrutiny on the one hand and increased pressure on public money and sponsorship, together with higher levels of accountability, on the other, will all place demands on museum boards and their chief executives for which few are at present equipped. People who run museums must pay more attention to the nature of museums as organizations. In many cases the governance, direction and management of these sophisticated institutions has outgrown not only the capabilities of the well-meaning amateurs who populate their boards but of directors and senior staff. A new quality of professionalism is now required. Without it museums will be unable to answer the crucial questions of long-term sustainability that most of them face.

The position of museum director—known as chief executive in some environments—which has always tended to be high profile is now becoming high risk too, partly because trustees are unclear about what they want and may not be sufficiently knowledgeable or experienced to know how to acquire those skills. In the United States in particular the dilemma between the need for a business head who is also capable of fundraising and a scholar curator who understands the collections and can articulate a vision for the institution seems impossible to resolve. Museum trustees must look at themselves before they determine the nature of the director they wish to appoint and the authorities that they wish to see exercised. In particular, they need to understand where the boundaries of executive and non-executive responsibility lie. All too frequently one sees directors spending their time flattering the egos of trustees or second-guessing their next move instead of running the place within the framework of a clear policy and mandate. Executive leadership in museums is a complex issue and good directors are hard to come by. Most governance models assume a single executive head for the organization. But alternatives have been tried, of which the celebrated division of responsibility at the Met in New York, between William Luers and Philippe de Montebello, was perhaps the best known. More recently the introduction of a Managing Director at the British Museum, to work in parallel with the Director, provides another example, although in this case the new structure was imposed from the outside by the Department of Culture, Media and Sport, the museum's principal source of funding.

Crucial to successful museum governance, direction and management is the definition and agreement of corporate goals. These are as essential in museums as in any corporate business environment. The concept has been difficult for some, and especially those who have thrived in the essentially sectarian structure of big museums in which their own ambitions, developed in carefully protected personal fiefdoms, could always take precedence. In the past some of the older tribal models appeared, at least superficially, to work, but this was in a less demanding age and such success as was achieved involved huge waste and duplication.[36] Today people are paid for what they do, not for what they are. Such is the multifaceted complexity of a modern museum, and the necessity to get the best out of limited resources, that even die-hard traditionalists are at last beginning to appreciate the benefits to all, including themselves, of new ways of working. The professionals in museums are beginning to recognize that they are all equals, each with an important contribution to make.[37] Senior management training, staff development programs, opportunities for directors and aspiring directors to acquire the necessary skills, will all contribute to radical improvements in the museum of the new millennium.

The museum of the future will be a vibrant place. It will transcend traditional cultural boundaries, embrace rather than exclude. Its licence from society to decode the past as well as the present stands every chance of being renewed. It will offer authority without being authoritarian, nurture scholarship and make its fruits widely available, look after its audiences as well as it looks after its collections. It is the eclecticism of the museum concept, its flexibility and adaptability that is so refreshing. The museum is becoming the ultimate medium of expression—of art, of history, of science. And it is worth our while remembering that museums share their ancestry not only with the *cabinet* but the penny peepshow. Others can trace their origins to the 150-year-old tradition of international exhibitions and world fairs, which, of course, also spawned today's theme parks.

Glenn Lowry, Director of New York's Museum of Modern Art, pictures the future of his museum as a scene from the Marx Brothers' *A Night at the Opera:* "a loud cacophonous environment in which fun is had by all."[38] And yet to go to MoMA today is to enter a cool, crisp and stylish place, not at all cacophonous, in which the E Type Jaguar and Bell 47D1[39] helicopter take their places as cult design objects with painting, sculpture and installations.

Victoria Newhouse recognizes this diversity:

> The public museum, which began with an educational impulse and later came to represent a new secular religion, is now widely perceived as a vehicle for entertainment. Recent decades have seen an immense increase in museum attendance, partly due to mass tourism, and crowded institutions have required adjustments in design to preserve their intrinsic qualities. . . . To think of art as entertainment is simply a return to the astonishment and delight associated with the first private Renaissance museums: a sensuous, thought-provoking discovery quite different from the dutiful didacticism of most large contemporary institutions, where visitors often spend more time reading about the art than looking at it. The museum's much-criticized shops and restaurants have the capacity, when handled in an appropriate manner, to serve this experience—just as jugglers, acrobats and other popular entertainers enlivened medieval religious festivals.[40]

In fact, the modern museum, irrespective of its theme or subject, through imaginative and professional planning, good design and programming is creating an extraordinarily persuasive and holistic dialectic in which an amalgam of collections, experiences and issues can live in happy and creative juxtaposition with exquisite food and drink and the best of retailing. As long ago as 1967 the Swedish artist Öyvind Fahlströhm (1928–1976) was convinced that museums would eventually ". . . involve theatres, discos, meditation grottoes, versions of Luna Park, gardens, restaurants, hotels, swimming pools and the sale of art replicas."[41] His "pleasure house" may well become tomorrow's museum model.

NOTES

1. The Dickinson Memorial Lecture was established by the Newcomen Society for the Study of the History of Engineering and Technology in memory of the late Henry Winram Dickinson (1870–1951), who had been a founder member of the society, its first Honorary Secretary, and, in 1932–3, President. Dickinson joined the staff of the South Kensington Museum—predecessor of the Science Museum—in 1895 and retired as Keeper of the Mechanical Engineering Department in 1930. Singer was also a founder member of the Newcomen Society, formed in 1920 following the James Watt Centenary Celebrations held in the previous year. The society meets in the Science Museum, London, where it also has its headquarters.
2. Charles Singer, "The Happy Scholar: The First Dickinson Memorial Lecture," *Transactions of the Newcomen Society*, Vol. 29 (1953–1954 and 1954–1955), pp. 125–132.
3. The Science Museum has its origins in the South Kensington Museum, established in 1857 on land acquired with profits from the Great Exhibition of 1851. The museum, which embraced both arts and sciences, divided at the end of the nineteenth century, the art collections forming the basis for the Victoria & Albert Museum, the science collections remaining on the original site on the west side of Exhibition Road. Construction of a purpose-built science museum began in 1913, was interrupted by the

First World War, and the building opened to the public in 1928.

4. Lyons, more than any other person, established the Science Museum in its modern form, as a national institution of international significance. He was Director from 1920 to 1933. His outstanding contribution to the museum is set out in David Follett, *The Rise of the Science Museum under Henry Lyons* (London: Science Museum, 1978).
5. As a generalization, the veracity of Singer's colloquialism is perhaps less open to incredulity than at first sight might appear to be the case. As a large cask for liquids, a hogshead was a measure of volume, of medieval origin, that held, variously, 63 old wine-gallons (equal to 52.5 imperial gallons), but for other commodities ranged in capacity from 100 to 140 gallons. An imperial gallon is today 4.546 liters. Popular myth would certainly support the contention that tea could be and frequently was consumed by British public servants in such quantities. Whether overseas guests could be induced to do the same is unlikely.
6. Jacqueline Trescott, "Exhibiting a New Enthusiasm," *The Washington Post* (21 June 1998), pp. A1, A16.
7. Jayne Merkel, "The Age of the Museum," *Oculus*, American Institute of Architects, New York Chapter (February 1998).
8. Ada Louise Huxtable, "Museums: Making It New," *The New York Review* (22 April 1999), pp. 10–15, reviewing *Towards a New Museum* by Victoria Newhouse (New York: Monacelli, 1999).
9. See Emlyn H. Koster, "In Search of Relevance: Science Centers as Innovators in the Evolution of Museums," *Dædalus: Journal of the American Academy of Arts and Sciences*, Vol. 128 (1999), pp. 277–296.
10. Neil Cossons, "The museums boom: When is a museum not a museum," *The Listener* (2 August 1984), pp. 13–17.
11. Neil Cossons, "The new museum movement in the United Kingdom," *Museum* (UNESCO) No. 138, Vol. 35 (1983), pp. 83–89.
12. See Victor T. C. Middleton, *New Visions for Independent Museums in the UK* (Singleton, Chichester: Association of Independent Museums, 1990); the same author, *New Visions for Museums in the 21st Century* (London: Association of Independent Museums, 1998).
13. Claus Kapplinger, "Architecture and the Marketing of Museums," *Museums International* (October/December 1997), pp. 6–9.
14. See Nicholas Powell, "Power building in Paris," *The Spectator* (22 August 1998), pp. 37–38, for an analysis of the complex political conventions that have regulated the world's most celebrated tradition of power building.
15. "Attractions in Stockholm by Stockholm Com" *www.stockholm.com/attractions.html.*
16. Caroline Reinhardt, "History with attitude: Elitism is out, populism is in," *The Spectator* (4 April 1998), pp. 43–44.
17. Dr. David Fleming, an articulate advocate of the museum as a vehicle for social history.
18. Heather MacDonald, "Revisionist lust: the Smithsonian today," *The New Criterion* (May 1997), pp. 17–31.
19. This international conference was held at the Society's House on 2 October 1990. The proceedings of the day may be read in "Scholarship in Museums," *RSA Journal*, Vol. 134, No. 5415 (February 1991) and include papers by Carlo Bertelli, Neil Cossons, Wolf-Dieter Dube, Michael Laclotte, Neil MacGregor and Michael Novacek.
20. In April 1984 the National Maritime Museum, Greenwich introduced admission charges for visitors, the first national museum in London to do so. By 1988 four others were also charging, resulting in bitter hostility from those that believed in free admission as a principle. Government at that time had a *laissez faire* policy, stating that charging was a matter for boards of trustees. The current Government believes in free admission and is implementing a program of phasing out charges accompanied by compensation for the lost income.
21. Calvin Tomkins, "The importance of being elitist," *The New Yorker* (24 November 1997), pp. 58–69.
22. Janet Daley, "The Mr Crapper approach to art," *The Independent* (25 October 1989).
23. To those who in the 1980s saw museums under threat from market force philistinism, this slogan, coined by the Victoria & Albert Museum's advertising agents, Saatchi and Saatchi, took on extraordinary significance; it symbolized their worst nightmares.
24. Dame Elizabeth Esteve-Coll was Director of the Victoria & Albert Museum from 1988 to 1995. She was relentlessly and brutally vilified by the arts establishment for challenging the dominion of curators and endeavoring to make the museum more widely accessible.
25. James M. Bradburne, "Dinosaurs and white elephants: the science center in the twenty-first century," *Public Understanding of Science*, Vol. 7 (1998), pp. 237–253. *newMetropolis*, the glamorous new science center in Amsterdam, designed by Renzo

Piano, has the distinction of falling into catastrophic financial difficulty within weeks of its opening.

26. Arthur P. Molella, "The Museum That Might Have Been: The Smithsonian's National Museum of Engineering and Industry," *Technology and Culture*, Vol. 32 (April 1991), pp. 237–263.
27. Frank Oppenheimer was the brother of the physicist J. Robert Oppenheimer (1904–1967), who directed the atomic bomb project at Los Alamos, New Mexico from 1943 to 1945.
28. See John G. Beetlestone, Colin H. Johnson, Melanie Quinn, and Harry White, "The Science Center Movement: contexts, practice, next challenges," *Public Understanding of Science*, Vol. 7, No. 1 (1998), pp. 5–26, for a review of the growth and current trends in science centers.
29. Follett, p. 55.
30. *Museums Journal*, Vol. 32 (April 1931), p. 39.
31. Quoted in Follett, p. 118.
32. See Neil Cossons, "A Perspective on the Nature of Industrial Collections," *Industrial Collections: Care and Conservation* [Proceedings of the conference hosted by the Council of Museums in Wales and the United Kingdom Institute of Conservation, Cardiff, 9–11 April 1997] (Cardiff, 1999), pp. 9–15.
33. Robert Bud, "Science, meaning and myth in the museum," *Public Understanding of Science*, Vol. 4 (1995), pp. 1–16.
34. See in particular Kai Bird and Lawrence Lifschultz, eds., *Hiroshima's Shadow: Writings on the Denial of History and the Smithsonian Controversy* (Stony Creek, Connecticut, 1997); Martin Harwit, *An Exhibit Denied: Lobbying the History of Enola Gay* (New York, 1996); Edward T. Linethal and Tom Engelhardt, eds., *History Wars: The Enola Gay and Other Battles for the American Past* (New York, 1996); and, Philip Nobile, ed., *Judgement at the Smithsonian* (New York, 1995), together with the reviews by William S. Pretzer and Otto Mayr in *Technology and Culture*, Vol. 39, No. 3 (July 1998), pp. 457–461 and 462–473.
35. The funding agreements between the department and the museums that it funds can be seen on its website: http://www.culture.gov.uk.
36. Some flavor of pre-reformation museum direction, albeit of a rather extreme example, can be gained from Roy Strong, *The Roy Strong Diaries: 1967–1987* (London, 1997). In a different context—London's Royal Opera House—the failure to understand the crucial role of the chief executive unfurls inexorably in Mary Allen's autobiographical *A House Divided* (London, 1998).
37. See Neil Cossons, "A New Professionalism" [Presidential Address to the Museums Association], *Proceedings*, pp. 1–2, Supplement to the *Museums Journal*, Vol. 82 (December 1982).
38. Quoted in Victoria Newhouse, *Towards a New Museum* (New York, 1999), p. 191, from Glenn Lowry, "Building the Future: Some Observations on Art, Architecture and the Museum of Modern Art," lecture at MoMA, 22 October 1997.
39. The Bell 47 helicopter was awarded the first commercial helicopter license, on 8 March 1946. It was designed by Arthur M. Young (1905–1995), poet, painter, cosmologist and author of *The Reflexive Universe* and *The Geometry of Meaning*.
40. Newhouse, p. 190.
41. Quoted by Newhouse (p. 191) from Annette Michelson, "The Museum World," *Arts Yearbook*, Vol. 9 (1967), pp. 84–92.

Communication of Science in the Deutsches Museum

IN SEARCH OF THE RIGHT FORMULA

Wolf Peter Fehlhammer

My entry into the vast and widely not understood field of popularizing science—throughout the text, technology is tacitly included—unavoidably is (how could it be otherwise?) that of a university teacher in chemistry who has carried out scientific research for twenty-five years. This explains the title and also why I felt particularly attracted by conference sessions on topics such as "Should Science Centers Do Scientific Research?" or "Novel Strategies to Keep Both Audience and Staff in Close Contact With Scientific Research," to which Jorge Wagensberg, also an active researcher apart from his duties as a museum director, had invited me in October 1997 to St. Louis[1] and in January 1999 to Calcutta.[2] Moreover, it explains the tenor of the following catalogue of strategies of growing impact on the visitor. "Steps to the ideal," so to speak. In my eyes, the primary goal is just to arouse a certain familiarity with science, its contents and products, yet also its nature, its way of reasoning, the way it works.

Proceeding on the assumption that we agree to make an exhibition on a scientific subject at all—which to begin with is by no means a matter of course, because a book or a lecture may be a much better source of information now and then, a fact constantly ignored only by museum people having an exhibitionist bent—the question of the contemporaneity of such exhibitions was clearly answered once and for all in the 1996 London Here & Now Conference held by the National Science Museum: It is a must![3]

When "Research" comes into play, Humboldt's ideal of the unity of research and teaching automatically makes sense, even in museums. However, we should perhaps be a bit less generous with the term "research" if it is only about counting visitors and the usual messing around with statistics. Restoration research, serious visitor studies and evaluations of exhibitions with the objective of possibly arriving at performance indicators, on the other hand, are as valuable and desirable as is genuine historical research, which in the Deutsches Museum ranks absolutely equal to that done in universities. The recent establishment of the Munich Center for the History of Science and Technology, with three university partner institutions and a lively cooperation with the Science Museum in London, the Smithsonian Institution in Washington and the Conservatoire des Arts et Métiers in Paris, adds to the overall reputation. Because 30 percent of our revenues are based on our research activities, scientific excellence is crucial for our economic survival.

In museums and science centers, surprisingly little is known about "experimental archeology." I was always fascinated by the interdisciplinary Plinius Project, for example,[4] and recently Otto Sibum has talked about another chapter in experimental history of science. To me, the reconstruction and revision of historical experiments and technologies, using original apparatus and contemporary methodology, seems particularly suited to attract the public.

I have talked about the necessity to keep both staff and audience in close contact with scientific research. Actually, on a moderate scale, I encourage my coworkers to lecture and to participate in research projects in universities. As a result of this *University-professors-as-director-curators* strategy, four members of the Deutsches Museum hold the title of professor because of their close links to universities. Only one of the professors is active in current experimental research, while the other three are in the field of history of science and technology.

But what about the audience? Well, we can rush to extremes such as: Every visitor should carry out actual research himself! Of course, scientists who have been in research for some time know this feeling that is beyond description, this moment of *Eureka*, when it works, when you all of a sudden understand what is going on, when crystals grow and you know you have a new compound and you are the first and only one in world who ever made it happen! That is the kick, that is the adventure of doing research, and many a scientist in charge of a museum (including me) wants his visitors to imagine what it is like. Still, it won't work, can't work, not really!

But then, where is the optimum on this continuum of ways of science communication that ranges from the boring to the unattainable? For my part, I declare it somewhere in between museum specific research on the one side, and cutting-edge scientific research carried out by the staff or the visitor (see above) on the other, which allows me to concentrate on what I call the mental, sensual and/or cultural approach that creates the very important emotions, the stimuli—not without pointing in passing, however, to what we think one should always keep in mind, particularly if one is building a Nobel Museum: That it is the scientist rather than the science that really inspires audiences. Therefore, an easy access to lecturers from universities and research institutions ("scientists to touch") is vital to introduce cutting-edge science in museums.

ART&SCIENCE—ART&TECHNOLOGY

Emotional stimuli are automatically provided by art. Up to the era of the Renaissance, art and science were considered to be just two sides of the same coin before they drifted apart, becoming the "Two Cultures."[5] Bringing them back together has almost become an obsession in today's cultural scene. Take, for example, the European Patent Organisation headquarters in Munich, which, as the President says, "very carefully makes its choice of the most allusive works to reunite science and art." Then, however, "Oppenheimer" finds himself somewhere in the office complex in the company of filing cabinets and non-flowering plants. I am talking of Christoph Bergmann's sculpture (Figure 1) named after the father of the Manhattan Project (Robert O.) and—very delicately—also of the science center movement (Frank O.).

But is this actually undoing the separation of art and science? I don't think so! And besides, it won't work, because it is counter-revolutionary! The parting of the ways of art and science, paralleled and followed by the separation of the disciplines, was the revolution and marks the awakening of modern science, whose success beyond all measure should actually have silenced the lamenting chorus of cultural critics and pessimists throughout the ages. It has not! Two hundred years after the first academies of science were called into being and the universities were finally brought up to date, we hear of a very similar accompaniment of the birth of museums: the opening up to the general public of the royal collections, their neat sorting into art and natural history, and after a due interval, science museums, a process admittedly not free of some pigeonholing, yet delightfully democratic.

Who really wants them back, those stuffy curiosity cabinets of the royal palaces? Actually, I am no stranger to private collectomania of *pre*-museum times, as it obviously still exists. Only recently we showed an exhibition on an unbelievable hotchpotch of furniture and art objects the mu-

FIGURE 1. Christoph Bergmann's "Oppenheimer."

Courtesy Deutsches Museum, München

seum had inherited from a well-to-do factory owner who had bought up his own world of dreams.

Now, however well-presented and atmospheric this exhibition may have been, it is certainly not what we strive for when talking of holistic approaches to contextual exhibitions. After all, we agree with Paolo Rossi that reviving ancient history is a futile venture.[6] What is there to be done, then, in order to meaningfully include art in science museums? A tentative answer is given by the formulation of a few (overlapping) criteria for and benefits that may be derived from the combination of art and science or art and technology, respectively, which we think range from being valuable to indispensable:

- Art must never be mere decoration!
- There must be an obvious, strong synonymity between the scientific or technological subject and the art installation from the beginning.
- Art creates immediate contextuality (providing an easy way to include, for example, societal issues).
- Art enhances the message (paintings on technical matters, *dioramas*).
- Art means a "sensual" ("cultural") approach (. . . "from the other side").
- Art can put science on trial, question its methods, point out its stereotypes, and so on.
- Art will appeal to and reach the number one target group of any museum, its *non-visitors*.
- Be aware of the esthetics of contemporary science and technology!

CONTEXTUALITY AND SOCIETAL ISSUES

Let me come back to "Oppenheimer," Christoph Bergmann's aluminum torso, which in its lower parts merges into the tail fin of an atomic bomb. Before it went to the European Patent Office, it was part of an exhibition in the aviation branch museum, where we experiment a lot with fine art. Here, in the vicinity of a jet plane, the sculpture has a most dramatic effect and leaves nobody cold. Endless associations arise. I mentioned the brothers Oppenheimer, the nuclear physicist, the hard science; of course, Dürrenmatt is in on it immediately, his exemplary border crossing in "Die Physiker" between the spheres of art and science, his maxim of responsibility; but there is also the deeply human educational concept of the founder of the San Francisco Exploratorium, mother of all science centers; and there is the tragedy of these individuals and, inextricably bound up with it, the tragedy of mankind.

For the museums, a recent alarming experience joins in which has its roots in the same American trauma: fifty years thereafter, the *Enola Gay* scandal tells us that interpretive social history exhibitions can be dangerous territory indeed. Today's Smithsonian publications reflect this new *horror controversiae* brought about by the neighborhood of the Republican capitol and the experiences with angry veterans, yet also caused by a chemical industry totally disappointed with the way curators appreciate its achievements.

This time it is a female torso (Figure 2) from which the bomb arises, Bergmann's way of expressing the ambivalence of technical innovation between progress and threat.

These are beautiful works of art opening up the dark sides of science and technology. With reference to the famous Otto Hahn-Tisch, and his staged experimental set-up for the discovery of nuclear fission, one of the museum's masterworks and myths,[7] we've come full circle.

Any criticism of the science-technology machinery may apply there and is legitimate, including the description of innocence lost once and for all.

ART&BRAIN—BRAIN SCIENCE: THE "SENSUAL" APPROACH

"The 90s have been declared the decade of brain; however, they seem to pass by without having solved

FIGURE 2. Christoph Bergmann's "Enola Gay."
Courtesy Deutsches Museum, München

the mysteries of our consciousness, our soul," writes Susanne Witzgall, curator in the Deutsches Museum Bonn, a *dépendance* that focuses on science and technology in Germany after 1945. And she continues: "Art approaches the object in quite a different way, now and then ignoring the scientific view of the world yet certainly questioning it. Thus attention is automatically drawn to philosophical, ethical and social aspects of scientific research, the dialogue between the two cultures is in full swing."

This program, which went into its third year, is produced in cooperation with experts from the research center Jülich, where brain science is a main area. Contemporary artists are invited twice a year to "decorate" a room of 5 by 4 meters. Among the last installations was Douglas Gordon's [original sound track: "I like the idea of introducing a popular myth in the context of popular science. Maybe there isn't so much difference between the two?"] "Life after life after life . . . ," centering on the legend of Störtebeker, the pirate, who after his execution walked headless past five or even eleven of his mates, thus saving them from the death penalty as negotiated with the town council beforehand. A headless dummy dressed in a T-shirt (which can [and should] be bought) is confronted with the close-up filmed face of Detlev Linke, a neurologist, who under the spine-chilling arrangement monotonously lectures on cerebral death, animal experiments and depth psychology (Figure 3).

FIGURE 3. Douglas Gordon's Art & Brain installation "Life after life after life . . ."
Courtesy Deutsches Museum, Bonn

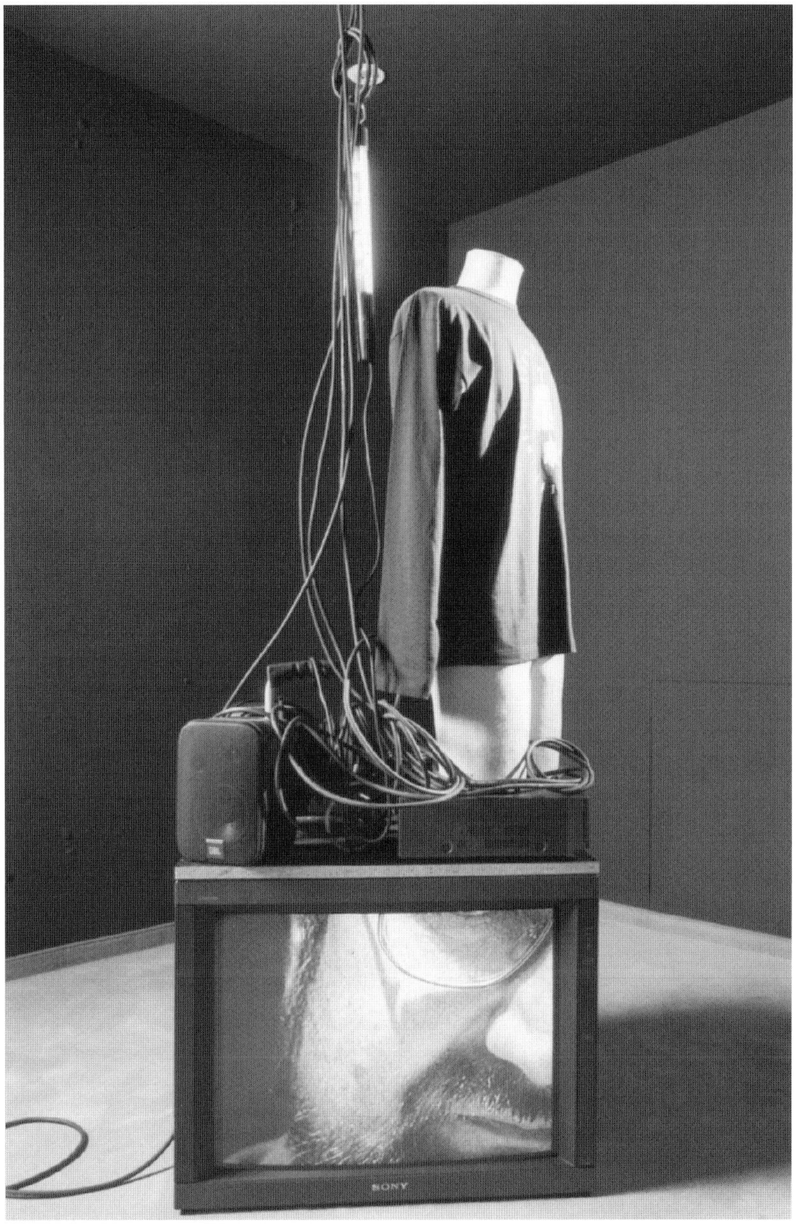

With Mark Dion's "Adventures in Comparative Neuroanatomy," an homage to the early collections in the corridors and cellars of scientific institutions and hospitals, an homage also to the vanished curiosity cabinets, the series ends for the time being. Dion particularly intends to reveal and thus again to question scientific methods and processes leading to knowledge, and he sets the visitor to thinking by displaying contemporary computer images of the brain, together with preserved specimens, in out-of-date systematics and untenable doctrines. None of these models can claim to represent the absolute truth. "Science museums and scientific collections mirror the theories of their epoch; they receive their *official story*," goes his message, "tinged with the philosophical convictions and ideologies of the respective time."

SCIENCE ON TRIAL

This leads directly to Theda Radtke's "Discovery of the 1000 species." She goes one decisive step further and hits the mark. On June 25–28 we had a series of events in the Deutsches Museum pretending to be scientific lectures, with all the ingredients deemed necessary to impress and convince an audience: the usual hour, the middle-aged professor who inspires confidence, the introducing host, the usual teaching aids—slides/transparencies—the stereotyped scientific talk of 50 minutes. Radtke had carefully studied all the rituals of a world existing far away from the art scene in so-called normality, whose esthetics *and* absurdity now became apparent through her initiative.

Visitors were introduced to the various forms of the 1000 species, the first of which had been discovered by chance; the lecturer went through collecting the species, sorting them into systems, their classification, their drawings (Figure 4), cross-breedings, analyzing properties, even granting of a patent (!)—that is, the whole process modeled after regular scientific research. It seems that the early natural history museums, already believed to be extinct just like the specimens they had studied, have been revived. But, actually, what has been demonstrated? This was left open. Real or not?—that was the question. It was the utmost in science theater. The resulting irritation—a hundred percent.

One thing is for sure, the audience has to be warned: Caution, art!

ART IN THE FLUGWERFT SCHLEIßHEIM

Meaningful, mutually enhancing associations of art&science and art&technology are indeed rewarding objectives which we particularly pursue in our aviation branch museum and the music gallery, an unusual but beloved addendum to a science museum. There, exhibition openings are regarded as vernissages that almost automatically attract the typical educated class or feuilleton people, interspersed with many a green-haired person who would never have set foot in a science center otherwise. Once they did, however, they got excited: Where else could Sabrina Hohmann have achieved a similarly strong effect? The staging of her work—eight small carpets showing pictures of bombed cities like Dresden or Cologne in bizarre metaphorical translation of the German word Bombenteppiche ("carpet bombing")—about military planes was absolutely congenial. Her gentle sorrow, her quiet lament were felt by the viewers. At the end of the exhibition, the eight carpets will have been worn out by the people walking on them, the pictures will have disappeared—just like the memories.

How loud was "Paul Klee," in comparison! Still, the 1997 temporary exhibition "Und ich flog" was the logical climax of the very consequent evolution of [what from that moment on I termed] Art& Technology or Art&Science in order to concretize and sharpen the vague KULTUR&TECHNIK, which for decades was more than the title of the Museum's journal, namely, its program.

Actually, the story was just too good and very much related to the locality: Klee actually was out there in 1916 dazzle-painting airplanes—imagine, all genuine Klees! There is also a mythical link with the Deutsches Museum, of which he must have been a real lover: the 18th-century wonderwork of the *Vogelzwitschermaschine* and a beautiful work of art modeled on it (Figures 5 and 6). But what is more: we, the engineers, the technicians could explain the art world for the first time, the meaning of the recurrent cryptic signs and symbols even in

FIGURE 4. Theda Radtke's "1000 Species."

Courtesy Deutsches Museum, München

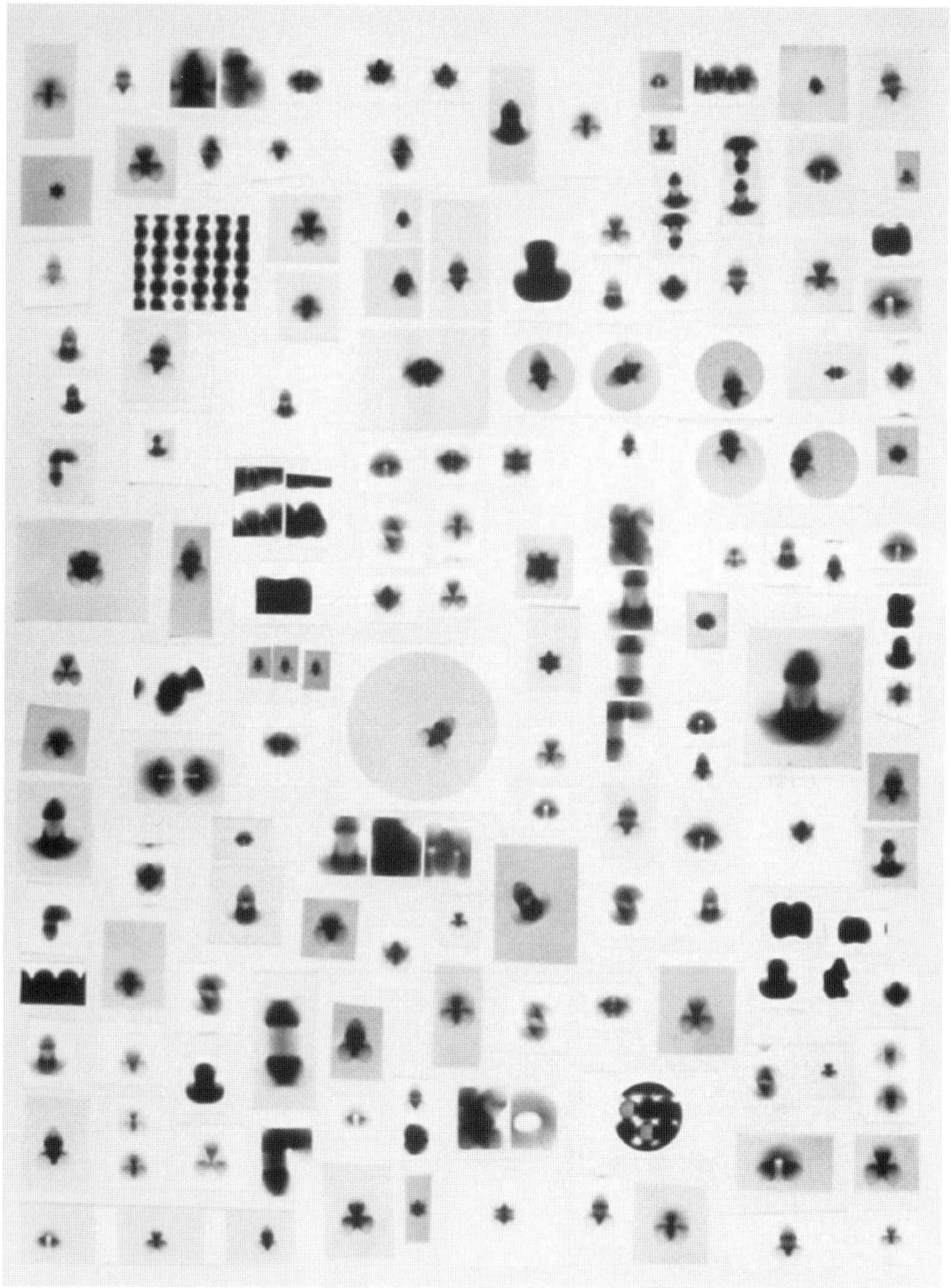

his later oeuvre. We could trace them back to real three-dimensional objects—kites, balloons, gliders, steel darts, crashing airplanes—that is, all the items of wartime experiences that must have hit him much harder than had been thought so far.

EMOTIONS WITHOUT LIMIT?

Fun, emotions, all right, but, of course, there are limits. A sharp line is drawn by a single word: Disney. It is used whenever museum people want to express their utmost disgust. On the other hand, one notices the overall appreciation of Disney's technical excellence and the perfect service it gives its customers.

Talking of border crossings, I would name "Körperwelten," an exhibition that recently made the news in Germany by attracting hundreds of thousands of viewers. You might say it was all exhibited dead bodies or remains thereof preserved by some polymer chemistry. However, it worked out, breathtakingly. The Mannheim museum of technology

FIGURE 5. **The "Vogelzwitschermaschine."**

Courtesy Deutsches Museum, München

FIGURE 6. **Paul Klee, Die Zwitschermaschine, 1922, 151 Ölfarbezeichnung und Aquarell, 41,3 x 30,5 cm; Museum of Modern Art, New York**

and labor, usually struggling along with 100,000 to 200,000 visitors per year, had 400,000 within two months, with people queuing day and night. This alone would make you think that a special nerve must have been struck, primordial fears and pleasures stirred up, a taboo broken.

THE BEAUTY ("DIRECT" ESTHETICS) OF CONTEMPORARY SCIENCE AND CULTURAL DISSIMILARITIES

I offer a final point. Let me put it this way: We don't need any art, science itself is sexy! Of course, science is also serious, difficult, impressive, sometimes boring, but, as a rule, it is of high repute, has its admirers, supporters, advocates, sponsors, and financiers. Obviously, theme parks feel the same way and are beginning to add some flavor of science half-heartedly. It has a fig leaf's function, because it makes them more respected, is something to take home. Science museums and centers move in the opposite direction: more fun, less learning. However, can we still be sure we are on the right track?

First, modern science can also be esthetic, even beautiful. Secondly, some (people), obviously, "like it tough (to read)." I choose two examples, both from our Bonn *dépendance*. The first is the Stellarator, a device that forces plasma into exactly the three-dimensional shape required for nuclear fusion experiments (Figure 7). So its beauty is purely functional, purely scientific. The other example is only two-dimensional, a page from von Klitzing's laboratory book, recording perhaps the moment of discovery of the Quantum Hall Effect, for which he received the Nobel Prize (Figure 8). Both objects stand for complicated science, very difficult to explain. Still, to our surprise, we watched people time and again lost in thought in front of this scribbling, impressed, if not in awe, observations which correspond nicely to what Renate Bader from the Free University of Berlin found out: that German students want complicated articles. Otherwise, they would not feel the information was sound. "Science in Germany is authoritative, not fun." This may be true not only in Germany but elsewhere.

As regards the beauty of von Klitzing's notes, we were so convinced of it that we had the notes blown up into a 15 by 20 meters poster, which is fluttering over Munich these days in competition with Andy

FIGURE 7. "Stellarator," a device for nuclear fusion experiments.

Courtesy Deutsches Museum, München

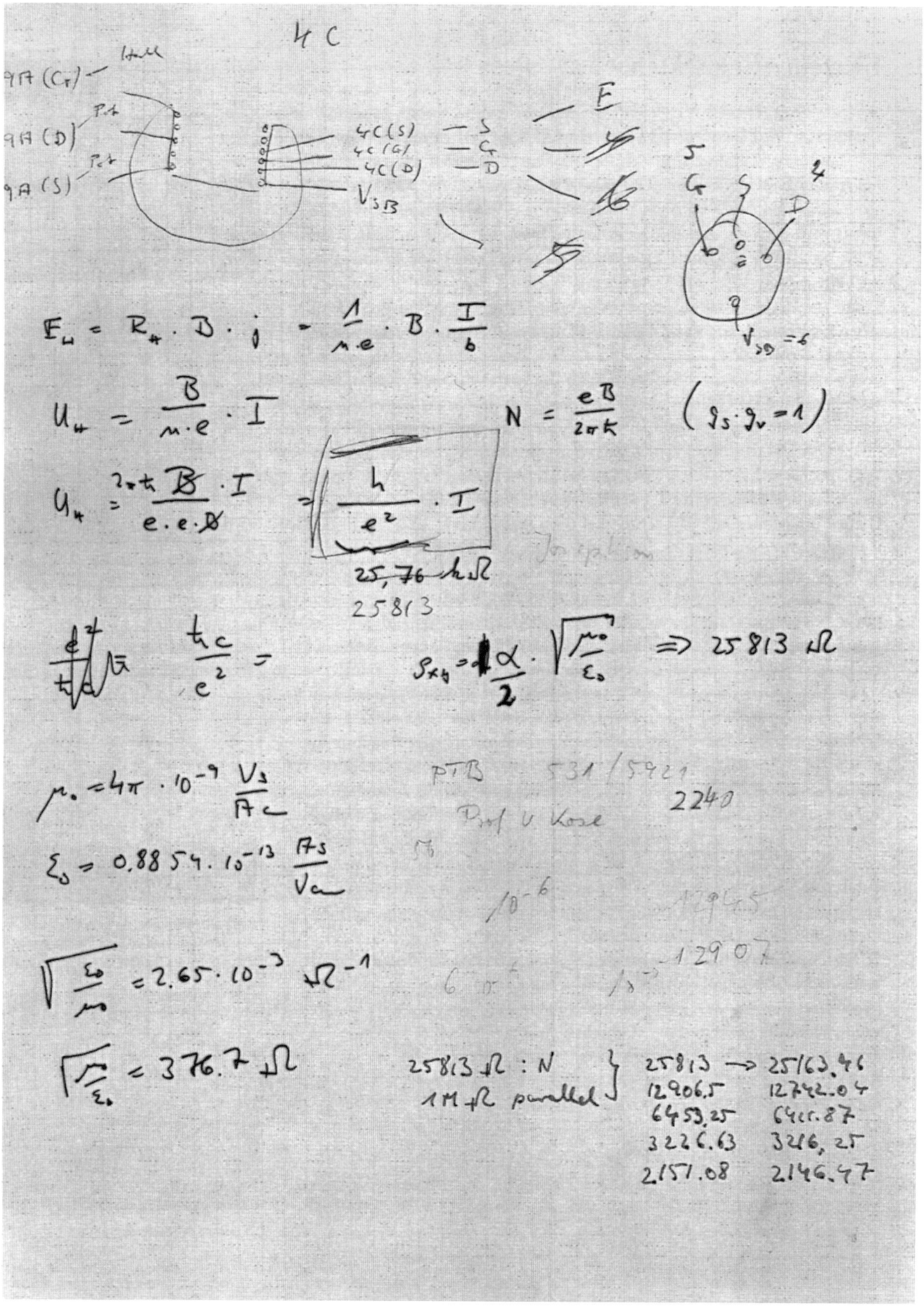

FIGURE 8. Klaus von Klitzing's discovery of the Quantum Hall Effect, for which he received the Nobel Prize.

Courtesy Deutsches Museum, Bonn

FIGURE 9. Blow-up of Klaus von Klitzing's notes.

Courtesy Deutsches Museum, München

Warhol paintings and sacral art (Figure 9). And it does its job: People wonder, stop, read again, and pass by shaking their heads. At times "public awareness of science" may start that way—with a Nobel prize!

REFERENCES

1. *Program of the ASTC Conference* (St. Louis, Oct. 1997).
2. *Program of the 2nd Science Center World Congress* (Calcutta, Jan. 1999).
3. *Here & Now Conference* (London, Nov. 1996).
4. Plinius (23/24-79), *Naturalis Historia*. (The work of the Plinius Project Group concentrates on metallurgy.) Series of Papers of the Georg-Agricola-Gesellschaft, Nr. 22, E.B.-Verlag (Hamburg, 1998).
5. C.P. Snow, *The Two Cultures*. London: Cambridge University Press, 1959.
6. Paolo Rossi, "Die Geburt der modernen Wissenschaft in Europa." Beck (1997): 62.
7. J. Teichmann, "Sozialgeschichte der Technik," eds. G. Bayerl, W. Weber. Waxmann (Münster, 1998): 199f.

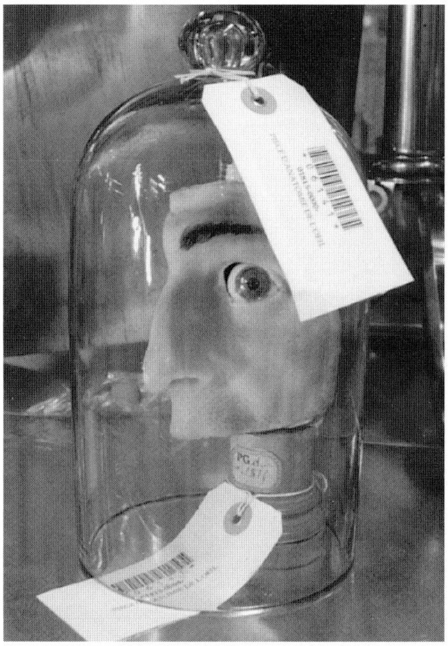

The Musée des Arts et Métiers

RENOVATION ISSUES (1988–1998)

Dominique Ferriot
Bruno Jacomy

The Musée des Arts et Métiers is a bicentennial institution, founded in 1794 by the Convention, at the request of Abbé Henri Grégoire, with the aim of "improving national industry." "In Paris, under the name Conservatoire des Arts et Métiers, and under the supervision of the Commission for Agriculture and the Arts, a depository will be created for machines, models, tools, drawings, descriptions, and books from all arts and trades. Originals of invented or improved instruments and machines will be deposited at the Conservatory . . . The Conservatoire des Arts et Métiers will comprise three demonstrators and a draftsman" (extracts from the decree of October 10, 1794).

The new establishment was set up in 1798 on the premises of the Abbey of Saint-Martin-des-Champs, and enjoyed the benefits of a unique location in a district of Paris bustling with artisans and small industries. The collection—hitherto scattered over a variety of locations—was gradually moved into exhibition galleries soon to be opened to the public. The draftsmen made the most of the installation period to copy the "insides" of machines and make detailed studies of mechanisms.

This short history serves as a useful reminder at the close of the 20th century, a time when the Conservatoire is engaged in a further transformation: the renovation of a museum rich with over 80,000 objects and 15,000 drawings and technical plans, and yet whose displays had aged so considerably that certain visitors had come to rank it in the "museum of museums" category. Despite its widespread renown, the Musée des Arts et Métiers was not a popular visiting site,[1] and had been left to die slowly, neglected by an institution that no longer understood the utility of a collection of now-historic machines in the development of contemporary scientific and technical training.

The project for the renovation of the Musée des Arts et Métiers—a key element in the State's "Grands Travaux" program of public works—is based around three observations. First, the importance of *experimentation* in scientific and technical culture and training for high school students, engineers and researchers; second, the irreplaceable role of the *object* as a tool in this active form of education, notwithstanding the versatility of audiovisual media and the vital importance of the written word; and third, the contribution of *history* to a better understanding of how technical innovations and know-how develop. With these factors in mind, the design of the new museum has been based on a thematic and chronological presentation, with the creation of workshops in each of the seven key "sectors" that structure the permanent display, and an opening onto the everyday context of technical

FIGURE 1. Archeological survey of the chapel of the museum (1993).

Photo: Philippe Plailly, CNAM, Eurelios

objects in spaces known as "kiosks," relating each subject to training, professions, and the Museum's partners in the industrial and research worlds.

In a museum, however, the public exhibition is only the tip of the iceberg. Much more goes on behind the scenes and, right at the heart of the system, is the *collection*, the key to the institution's legitimacy and richness. Not all publics can access the "reserve collections" that form the core of a museum. The creation of "visitable" reserve collections—in other words, reserves that are accessible to anyone conducting a research project—was the first step in the renovation of the Musée des Arts et Métiers.

VISITABLE RESERVES

What curious visitor has never dreamed of entering into the reserve collection of a museum, often envisioned as a romantic hotchpotch of busts of inventors, forgotten machines and the—sometimes antiquated—traces of a prolific past. This image does not correspond to reality, since curators are by vocation people who classify, group together, and draw up inventories. The *computerized inventory* of the collection at the Musée des Arts et Métiers has been the basis of the museum design project. It has made it possible to transfer the objects in an orderly manner to the new Reserve store created in Saint-Denis (5 kilometers from the Museum's display site in the rue Saint-Martin, in Paris). The building, designed by architect François Deslaugiers, is in two parts: the two-level wooden "case" in which objects from the collection are stored on rows of racking bays, with appropriate signage, and the "life" building, with its steel fuselage, containing the restoration and photography workshops, the quarantine room, and the research offices reserved for researchers authorized to work in the Reserve store.

"Visitable" by this specialist public, the Reserve store is a large, airy site in which the object or series of objects being researched can be viewed. Work can be facilitated by prior use of the Museum's objects database that can be accessed via the Internet (http://www.cnam.fr/museum/) and is listed in various publications, notably *L'Album*,[2] a CD-ROM presenting a journey through four hundred objects in the collection, both documented and animated.

"All in all, the immense world of objects reveals more about the mind than the mind itself."[3] This remark, made by the philosopher François Dagognet, is the key to the extensive research still to be carried out on artifacts deposited at the Conservatoire over a period of more than two hundred years. The only collection of its kind in the world, the collection at the Musée des Arts et Métiers comprises some of the machines and instruments that founded modern experimental science: original demonstration machines used in 18th-century

FIGURE 2. The Foucault pendulum (1855).
Photo: Jean-Claude Wetzel, CNAM

physics cabinets; weighing scales, gasometers, and calorimeters made by Lavoisier, Fortin and Meusnier and used for experiments on water synthesis or metal combustion; the first automobile or flying machines: Cugnot's "Fardier," Ader's airplane, models of trains, velocipedes; the first programming machines (Pascal's calculating machines, Hollerith's machine, the first computers); symbolic objects later copied by all technical museums (Foucault's pendulum, machines for producing electricity, automaton-musicians . . .). From among these objects and sets of objects (weights and measures, batteries, models of steam engines, glass and ceramics collections . . .), a selection of "flagship" artifacts and groups of objects of particular significance will be put on display to the general public. Nine-tenths of the current collection will therefore be kept in the Reserve store, which will also house essential objects testifying to our contemporary technical heritage.

The success of the symposium organized at the CNAM in September 1994 on *The Reserve Collection in Museums*[4] is easy to understand. The Reserve store has a vital role to play. It acts as the "lungs" of the museum, the focal point of a much-needed partnership with the industrial and research worlds, and a center for the conservation of an increasingly active "memory" of the past. But what should be conserved? For what purposes? With what means? These questions—asked daily by the curator—are even more pertinent in a technical museum, given the rapid obsolescence of artifacts and the difficulty encountered in conserving objects made to be used, worn out or transformed. The future of science and technical museums therefore depends to a large extent on their ability to make reserve stores a living place open to researchers, while remaining a center for conservation, restoration and maintenance, equipped with state-of-the-art facilities and the protective

FIGURE 3. The reserve collection of the Musée des Arts et Métiers, Paris.
Photo: Pascal Dolémieux, Métis

measures vital to an effective management of the collections.

The renovation of the Musée des Arts et Métiers will be judged as much on the successful use of its Reserve stores as on the hoped-for success of the new displays to be put on show to the public.

THE EXHIBITION: ON THE TRAIL OF TECHNICAL INNOVATION

The renovation of a bicentennial museum has very little in common with the creation of an entirely new establishment. The existing elements, witnesses to the history of the site and its collections, must imperatively be taken into account in the overall project, and especially in the general organizational blueprint for the future Museum.

The museum as part of a history

In the case of the Musée des Arts et Métiers, the above constraints—which have been evident since the launch of the project—are present on three main levels: the *collections*, the *site* and the *evolution of techniques*. What is so special about these collections, founded in the mid-18th century with Vaucanson's mechanics cabinet, is that they describe, as accurately as is possible, *the contemporary state of technical science* in each of the various periods concerned. Within the context of the foundation of the Conservatoire, the collections reflected the overwhelming desire to place the site within the economic context of the nascent industrial revolution. This called for the reproduction—either life-size or in the form of models—of the machines, tools and instruments that favored technical innovation. Replicas of British textile machines, new machine-tools, and drawings of new processes were intended to inspire the public at the newly-created Conservatoire—artisans, technicians, and also curious observers—to come up with original solutions and, on a broader level, to acquire the *technical culture* which we in turn aim to stimulate in the modern-day visitor.

A second extremely limiting element was the site. Over time, the institution—devoted to the collection, demonstration and teaching of a veritable culture of technical modernity—turned into a museum frequented by people familiar with technical science, or with a passion for a long-lost industrial heritage. Not only the objects, but the display cases, the chapel, indeed the very walls of the old abbey have become active elements in a place that is truly unique; they must not be neglected at a time when the public displays are being renovated.

The third key element is the evolution of technical science. The two constraints expounded above pose a prickly problem in terms of exhibition design: how to stay loyal to the original desire to present technical innovation when, as a direct result of this, the site has become an unrivaled temple to bygone techniques? The third constraint only complicates the problem. Over the past two hundred years, the objects and walls have stayed in place, the collections have accumulated, and the institution has become both a museum *and* a higher education establishment. Above all, in parallel to this, the world has changed immensely.

Following on from the industrial revolution that grew out of the invention of the steam engine in the late 18th century, other upheavals were to come: those provoked by the onset of electricity, then electronics and computing, synthetic materials and reinforced concrete, television and nuclear power. In the 1920s, at a time when the Conservatoire was gradually affirming its vocation as a higher education establishment focused on the promotion of work, the Museum suspended its evolution, and became frozen in time, displaying inestimable collections that were increasingly remote from the modern-day reality of technical innovation.

Over a period of almost half a century, several Museum officials made attempts to renovate rooms, developing the sections devoted to audio-visual technology and everyday life, for example. But beyond the structural problems involved—and in particular the institution's lack of interest in its historical collections—was the fact that the techniques themselves were becoming increasingly *immaterial*. While machine-tools, looms and vehicles could still be used effectively for demonstrations, it was becoming more and more difficult to explain the new techniques of the 20th century using animated machines or models. How can one convey the operation of a radio receiver, a radar, a computer, or even a turbojet, without being forced to resort to

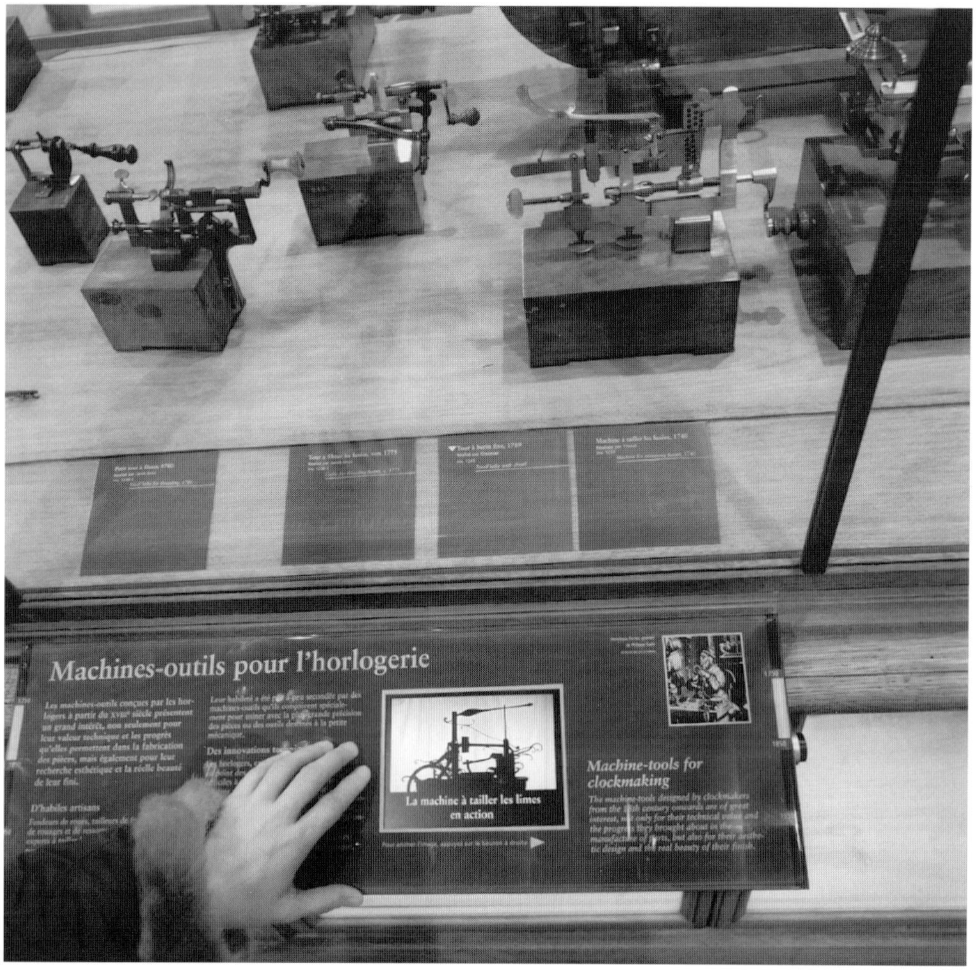

FIGURE 4. Animated captions for flagship objects in the new permanent exhibition, Musée des Arts et Métiers, Paris.
Photo: Pascal Dolémieux, Métis

the *basic blueprint*, the model that would now be called virtual?

This evolution in "modern" techniques must be taken into account in the renovation project. The steering report drawn up by Pierre Piganiol in 1989[5] clearly stated that there was no question of making up for the existing "gap" in the collections as concerns the period from the inter-war years to the present day. However, it would obviously be unthinkable, as part of an overall policy of revitalizing technical culture, to rebuild a Museum whose presentations stopped—in the case of many fields of technical science—with the onset of the consumer society.

The choice of the future layout of displays therefore had to be based on the existing situation—which is extremely limiting—and led to the general organization of the museum into seven main sectors. In effect, the existing displays, which had been structured over time into more than 25 sections often linked to scientific disciplines (optics, acoustics . . .), technical fields (mechanics, electricity) or industrial sectors (transport, clockmaking), no longer corresponded either to the state of technical science in the late 20th century, or to a global and reasoned understanding of the evolution of technical science in our western society.

As a result, we needed, first of all, to give a new sense of coherence to a collection which is in many respects disparate and, secondly, to help a public generally uninformed as to the history of technical science to acquire bases, indispensable guidelines for apprehending this key element in our understanding of the world.

Choosing the "main sectors"

The new structure of the permanent display is therefore the product of all these constraints; its various components are sometimes classic, sometimes original.

Three sectors belong to the former category and are part of a certain historical continuity: they are Transport, Mechanics and Energy. Already very prominent in the history of the Conservatoire and its collections, these areas are necessary elements in a "classic" visit of a technical museum. They exist in most other museums, usually more recent than the Conservatoire but which grew out of the major transformations of the 19th century.

- *Energy*, one of the foundations of the industrial revolution, is mainly dealt with from the standpoint of energy "converters": steam engines, generators and electric motors, internal combustion engines;
- *Mechanics* takes us back to the very origins of the collection, with Vaucanson's machine cabinet at the Hôtel de Mortagne. It is, in a sense, the backbone of the Museum;
- *Transport*—featuring a number of spectacular objects, from Cugnot's Fardier to turbojets—shows the incredible impact these technologies have had on our everyday lives, from the introduction of railways to the conquest of space.

The four other sectors, unlike those listed above, are completely new to the Museum. They represent either the development of a whole new field, or an effort to shed new light on collections neglected until now. The *Building* section is in the latter category.

Those collections relating to civil engineering, highways, and public works are totally unfamiliar to many visitors, including some of the Museum's "regulars." And yet they constitute a group of objects of rare quality. We have chosen to highlight this aspect because, on the one hand, these collections are more indicative of construction *techniques* than of architecture—structures, calculations and scaffolding, for example, or what generally used to be termed "the art of building"—and on the other hand, because no other collections of this type are on public display in France. Some of them have been in the Parisian Museum of Public Works or on loan to foreign museums that attribute a more important place to this theme—notably the Deutsches Museum in Munich. In France, there was a major gap to be filled, particularly in showing how the evolution of building techniques over the past two centuries can be traced through the use of iron, glass and concrete in our everyday environment.

Unlike the Building sector, a large part of the collections presented in the *Communications* section were on display in the Museum's old rooms. The main problem with this vast field lay in determining its limits. As a result, one of the largest spaces in the Museum—the first floor of the Vaucanson wing—was chosen to present the various forms of communication (apart from transport, of course), from printing to telematics. At the beginning of this century, it was still quite conceivable to present images, sound, the telephone and radio, for example, in several separate rooms. At the turn of the 21st century, however, such an approach would be quite impossible, due to the extraordinary upheavals that have taken place within this sector since the inter-war period.

How, for example, could we convey to visitors the profound *interweaving* and cross-fertilization between the telecommunications and information-technology sectors as a result of electronics technology—notably since the invention of the triode—without covering this vast field of communications both historically and *globally*? As with the other sectors, this involved inevitable redundancies and often heart-rendering choices, especially in the case of the computer, dealt with in this section in terms of the *technologies* involved in information technology (hardware and software), whereas its *use* will be evoked in most other sectors in illustrating the contemporary era.

Scientific Instruments and *Materials*—surely the most original of the new sectors—group together a number of sections previously either dispersed throughout the Museum—such as optics, physics and acoustics (Instruments)—or absent from the public exhibitions for many years, such as industrial chemistry or metallurgy (Materials).

The idea of grouping together all these tools and instruments used in the development of sciences under the label "Scientific Instruments" was immediately endorsed by the Scientific Committee for the renovation of the Museum and, in particular, by our colleagues abroad, since the inestimable richness of our scientific collections has won renown well beyond our national frontiers. The creation of a global and historical "dynamic current" running throughout these disciplines—presently compartmentalized in school and university curricula—will encourage a better understanding of the development of science over a period of more than three centuries, and of the importance of instrumentation—and, of course, of the designers and makers of these instruments—in the progression of science.

Finally, the decision to devote a whole sector to *Materials* was dictated both by the *contemporary state* of the technologies involved and by the importance within this context of the nascent industrial revolution. The latter was born in 18th-century Britain, and was notably characterized by profound transformations in the fields of energy and materials. As a result, the rooms devoted to energy and the steam engine will be given a prominent place. Following on from the water-mill, the steam engine brought not only a sizable increase in the available power, but also wide-scale geographical redistribution of the centers of industrial production. Indeed, manufacturing developed most extensively in sectors relating to materials: the development of iron and glass, and industrial chemistry, for example. The decision was taken to display all these materials together in one group precisely because they have been closely linked over time: not only in the 18th and 19th centuries, but in the period since World War II as well. What some may call a "hyperchoice of materials" corresponds to the intrinsic ties between synthetic materials, biology, chemistry, metallurgy, and the treatment of wood and ceramics, to name but a few. One only has to observe the way a modern car is built to assess the progress made over the past century in the materials field.

Following the thread of time

Since a museum is, by definition, an institution founded on memory, time is a crucial factor. However, it is curious to note that the way in which time intervenes can vary considerably depending on the type of museum. In ethnographic museums, geographical zones form the basic framework, and time plays a minimal role. In art museums, the types of materials or media used in the works themselves—sculptures, paintings and drawings—are frequently combined with a division into major historical periods—prehistory, Ancient Egypt, medieval sculpture, romanticism . . . History is omnipresent via the artistic currents that run through the displays.

In science or technical museums, the chronological framework is generally set aside, and two reasons may be given for this. In the case of natural history museums, the different disciplines structure the displays—mineralogy, botany or paleontology, for example—and the time factor as it relates to species, rocks or the earth, covering an extremely long period, is given a secondary role, at least until the recent creation of the Main Gallery at the Muséum d'Histoire Naturelle in Paris, since *evolution* is the fundamental principle behind this recently renovated space.

Technical museums, for their part, tend to view the temporal dimension as a secondary issue, generally because they were designed as state-of-the-art museums, like the ancestor to them all: the Conservatoire des Arts et Métiers. Although they have gradually developed historical collections, the demonstration of the technologies involved in machines and systems—the "how it works" aspect—was for a long time the main mission of museums of this kind.

As the years went by, this original and deliberate setting-aside of history came to be seen as a failing, one which certain museums made up for by reintroducing historical presentations of technologies, providing the public with more familiar reference points. This is the case with the Science Museum in London, which has designed a permanent exhibition space entitled "Synopsis, an introduction to the Museum," as a prelude to the various thematic sections of the museum.

In art museums, time references remain vague for most visitors, and yet they do exist, since art history is covered—although on a limited scale—in traditional history teaching. The same does not apply to technical science and the arts and trades. Very few people are aware of the importance of technical innovation in the history of our societies. This fact has translated into the need to provide visitors at the new Musée des Arts et Métiers with a chronological framework, helping them to situate the technologies on display in relation to their own historical reference points. Each of the seven sectors in the Museum will therefore be treated chronologically, from its origins to the present day, or from the present day to its origins if the visitor chooses to follow a trail moving back over time. This chronological framework, along with the general thematic blueprint will, for the first time in the Museum's history, allow an understanding of the dynamics of technical change—the upheavals generated by the onset of steam power, the metric system or electronic tubes, for example—and also the prevailing "trends" over time within a given sector: the 18th-century physics cabinets in scientific instrumentation, or the birth of talking movies in Communications, for example.

This presents a problem, however. As soon as one attempts to "recount" any story—technology in Western societies, for example—the inevitable problem of *reference points* comes to light. Just as a book is structured into chapters, an exhibition must be divided into periods. And it is here that we encountered a major stumbling block in choosing dates or events to structure each of the sectors.

Since we were determined to create a museum that is both coherent and unique, rather than a series of seven independent units, we chose the same reference points for *all* the sectors. Needless to say, landmark events in the history of scientific instrumentation do not coincide with those that marked the energy or mechanics sectors, and the key dates in technical history generally correspond even less to political, economic, social or art history. This matter has been the subject of countless debates within the Museum, the Scientific Committee, and during the various meetings that have punctuated this epic renovation project. The solution finally adopted was to take "round-figure" reference points: dates that are easy to remember—1750, 1850 and 1950—possibly corresponding in several cases, although not necessarily, to major events that we will not describe in detail here.

These dates divide each of the sectors into four eras. Landmark dates are not opaque doors, but approximate limits marking out "frontier regions."

In short, the theoretical framework of the future Museum can be expressed as a synoptic table constructed around a number of themes and eras, allowing visitors to find their bearings within the general plan, wherever they may be located in the building.

The "open spaces," technology in perspective

It may seem contradictory to want to express history from a *global* point of view, and yet structure the museum into seven separate sectors. That contradiction certainly exists, but the field of knowledge the Musée des Arts et Métiers is dealing with is much too vast to be taken all in one piece within a uniquely historical framework. With a view to overcoming this problem, a number of intermediary spaces have been designed. These spaces—located at the crossroads of the main sectors—are called in French *places*, or "squares," by analogy with the urban environment in which the transfer from one street to another, from one district to another, involves crossing a square: a resting place with benches, a breathing space with trees or a fountain, a place for finding one's way with maps and access roads and, finally, a meeting place with areas to stroll around in.

The open spaces in the future Museum will not have quite so many functions, given the particular configuration of the site, and the metaphor is therefore only a partial one. However, they will play two main roles: as a *breathing* area between two sectors whose centers of interest may be very different and, above all, as a place for *putting into perspective* the various themes distributed throughout the seven sectors.

The open spaces will show that, beyond the technical and industrial sectors, there exist a number of historical constants, very strong links uniting apparently disconnected worlds. *Breakdowns and Accidents*, for example, will be one of the themes tackled in the five open spaces. With the help of a few objects and a short audio-visual presentation, visitors will be able to measure the importance of breakdowns in all economic, technical and industrial sectors, throughout all periods of history. Technical progress has consistently learned from failures and errors that sometimes produced accidents: boilers exploding, cars breaking down, computer bugs; every technical system comes with its own set of dysfunctions that engineers and technicians must be able to remedy.

The *Renaissance* open space offers a global view of science and technology during a particularly prolific period of innovation in the fields of mechanics, energy, and architecture. The open space devoted to *Drawings and Models* will shed light on the way machines have been represented over time: from the engineer's notebook to industrial drawing, from the animated model to virtual modeling.

Five such open spaces will be distributed along the visiting trail. Since one of the roles of the open spaces is to provide a breathing space, it is important that the exhibition design in these intermediary, transverse areas should contrast sharply with the style used in the sectors themselves. It must be clear to visitors that, for example, they have just left the Building rooms and will soon be entering the Communications rooms. The topography of the exhibition spaces is limited by the linear path followed through the Museum, but this does present the advantage of avoiding hybrid trails which, in some museums and centers, can lead visitors to move unwittingly from one space to another, and from one theme to another.

The exhibition design in each sector is therefore primarily based on *objects*, whereas the emphasis in the open spaces is on *discourses*. In one case, the visitor observes the artifact, and can then go and seek the information he wants; in the other, he watches a short presentation that arouses his curiosity, then he looks at the objects on display as an illustration of the discourse he has just seen presented. These spaces are very limited in surface area (50 to 80 m^2), but their "dramatic" intensity and "discourse" approach will make the open spaces an innovative element in exhibition design.

In the case of the *Machine Man* exhibition presented in October 1994 and January 1995 in the ambulatory of the Museum chapel, an eight-minute audio-visual show entitled "Vaucanson's Dream" closed the visit by plunging visitors into the world of automaton creators and robot makers, and encouraged them to reflect on artificial beings. The show, created by Bernard Szajner, combined real objects—a lathe, a loom, and various Vaucanson machines—and virtual images, using state-of-the-art audio-visual and multimedia technologies.

"Vaucanson's Dream" was not an "open space" as such, but the show—which made a strong impression on visitors and brought the main theme of the temporary exhibition clearly into focus[6]—gives an idea of the exhibition-design approach to be developed in these areas.

A priority trail

The "open spaces" and "main sectors"—the two key elements in our approach to museum design—create a consistent pattern that runs throughout the trail. A visit of the entire Museum—while it can only be superficial within a reasonable period of time—will therefore proceed as follows. From the reception hall, visitors will take an elevator that transports them directly to the second floor, at the entrance to the Scientific Instruments sector. Here, they will discover the oldest, and sometimes most prestigious instruments and devices on display in the roof area which, until now, was used for storing the reserve collection, and was therefore non-accessible to visitors. This desire to let the public experience the delights of discovering the collections in this superb wooden "casket" was an original idea of Andrea Bruno, the architect chosen to design the renovated Museum. His aim was to share the intense emotion he himself experienced on his first visit to the Conservatoire.

The second-floor visit then continues into the rooms that were used for displays of weights and measures in the 19th century, but had since been turned into reserve stores. At present, this area comprises collections of scientific instruments, followed by the Renaissance open space and the Materials sector. A staircase leads down to the first floor, where visitors can stroll around the Vaucanson courtyard and discover, one after another, the "Drawings and Models" open space, and the Building, Communications, Energy, and Mechanics sectors, interspersed with open spaces entitled "Technologies Outside Europe" and "Game and Technology," and ending with "Breakdowns and Accidents."

Taking the main staircase, visitors return to the ground floor, where they discover the Transport sector, and then end the visit with the chapel, now presented as a "lieu d'émerveillement"—"place of wonder." It is mainly filled with transportation devices (most of them very large), including aircraft, omnibuses and automobiles.

This general visit is the main trail visitors are

FIGURE 5. **Musée des Arts et Métiers permanent exhibition.**

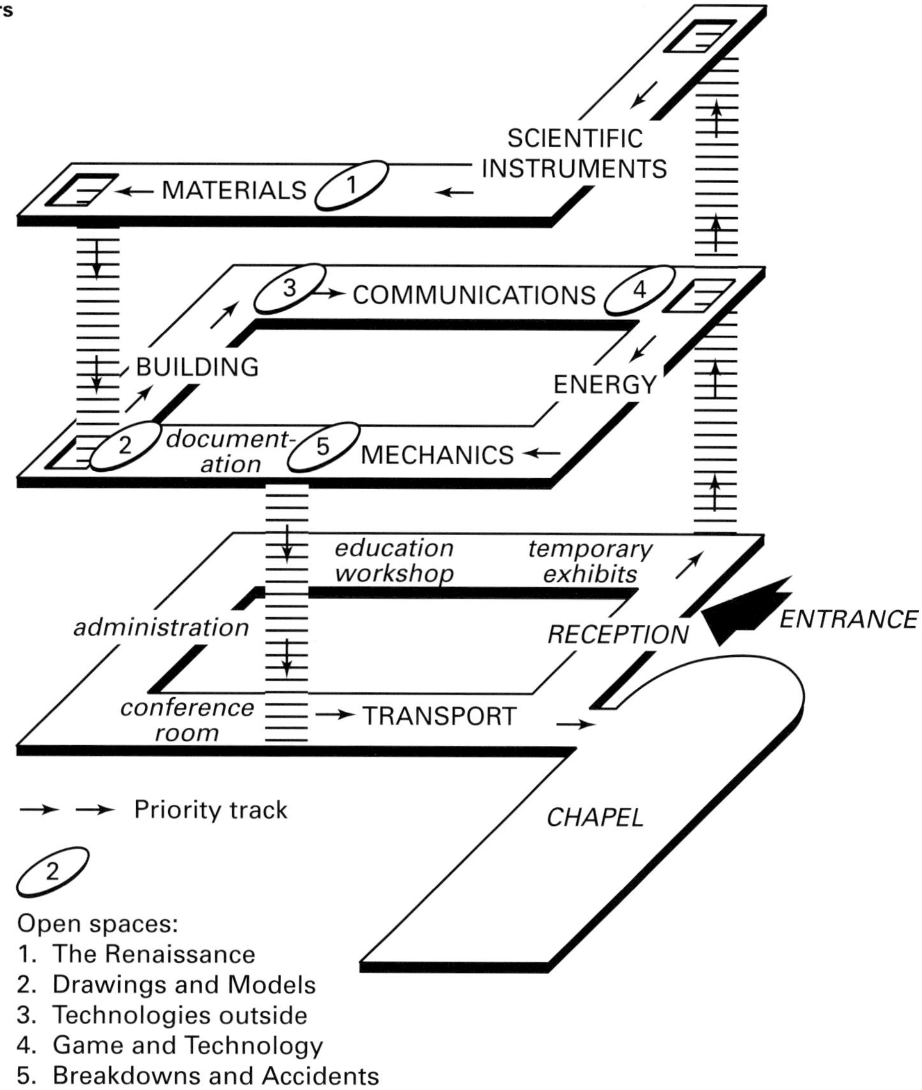

advised to take, although a number of "ball-and-socket" alternatives are possible. For example, a rapid trail for younger visitors would consist in visiting the Energy, Mechanics and Transport sectors and the chapel—areas featuring more animated displays, and the ones most frequently requested by younger audiences. Taking a different approach, it is possible to short-circuit the first floor and visit only the Scientific Instruments, Materials and Transport sectors. Whatever the case, the chapel, the historical heart of the Museum, with its prestigious nave and the presentation of the history of the site in its ambulatory, can always be accessed from the reception hall by visitors with a particular interest in the sensitive and spectacular approach it offers.

Hierarchized displays

While the "open spaces" focus on the "discourse" element rather than objects, in the vast majority of the Museum—over nine-tenths of its surface area—priority is given to the collection. Even when they are emotionally charged, however, the objects on display are often difficult to access. In an old physics room, for example, how could essential objects such as Nairne's electrostatic machine or one of Coulomb's original balances be made to stand out among hundreds of similar objects arranged in rows in display cases? Without the help of a demonstrator or specialist, the visitor walking through the Museum would be incapable of distinguishing objects of special interest within this huge mass. Given the intention that the renovated Museum should not address a select public of historians or collectors, a means had to be found to "hierarchize" displays, to guide visitors through its rich collections.

The museum can be visited on two levels, and these are explained to visitors as soon as they arrive :

- A *"flagship objects" trail*, structured around exemplary objects. Each object is presented with information highlighting its importance in the evolution of technical science and its consequences for society. The main theme of the sector in question is developed, helping visitors to understand it more clearly.
- A *"corpus" trail*, providing a comprehensive visit of the sector, immersing the visitor in the richness of the collections, and featuring a large number of objects grouped into thematic units that complement the "flagship objects" and give greater depth and breadth to the main theme of the sector.

Throughout the trail, flagship objects are indicated by special signs, a sort of lectern providing more detailed explanations than those alongside the other objects. This means that visitors can either make a rapid, summarized visit, passing from one flagship object to another, or appreciate the full richness and diversity of the collections, wandering through according to their particular interests and the time available.

One particular category of objects will be on display at all levels of the exhibition: the *series*. These are essential in understanding the evolution of objects and the way technical thought progressed over time. They include series of batteries, gearing systems and electronic tubes, grouped together into logical sets. They have been designed to make visitors aware either of the variety of technical solutions imagined for a given problem, or of the different stages in the creative process for a given object, over a period that may be long (a century of trial and error leading to Watt's steam engine), or short (the twenty years it took for the modern bicycle to develop). The series concept, treated here as a specific way of presenting objects, provides a basis for understanding the principles of machine development.

Means of access to information

While priority is always given to the object, visitors obviously expect to be provided with information as to how the objects work, their history, and their place in the history of our society.

Throughout the exhibition, information will be presented to visitors in two ways. The first—the *demonstration*—is rooted in the very history of the institution, and is still the method of communication most suited to the transmission of technical know-how. Demonstrators will be present on-site, either to set the objects in motion in their display cases, or to animate a session in the *sector workshops*. Each of these sectors will therefore comprise an educational "pole" featuring animated models that can be handled directly, and also a computer workstation linked to the Museum's database, in which visitors can consult all collections relating to the sector in question, in the form of texts or fixed and animated images. In this area, demonstrators will provide regular explanations of the principles and functioning of the machines and instruments, placing them within a dynamic historical context.

In addition to this, "self-service" access to information will be available in the form of specialized workstations located all along the trail, on the outer edges of each room. These *Albums*, as they are called, can be "leafed through" by visitors, and will be based around an event in the history of the sector (drilling the Suez Canal, the first cinema session by the Lumière brothers), reflecting the political, economic or artistic context of the period in question. Large touch screens known as *postes histoires* ("history stations") will shed even greater light in this respect by evoking a theme that has recurred throughout the history of the sector, or one that is linked to two separate sectors. Navigating with the help of hypertext, the visitor will discover, one by one, using texts, animated diagrams and period films, the role of scientific-instrument makers in the 19th and 20th centuries, or the role of materials in transportation, for example.

Finally, the *postes personnages* ("personality stations") will offer a more sensitive approach to the men and women who have played a role in this great story. A number of spoken quotations and commentaries concerning inventors and technicians—not all of whom are familiar names to the general public—will complete this global approach to the sector and enhance knowledge of the objects on display. While the history stations and albums (about three of each per sector) can be consulted sitting down on benches arranged next to the windows, the personality stations are located near the sector workshops, acting as a complement to the central "educational" structure.

These special facilities for public use, combined with the *kiosks* encountered further on in the visit, represent a real design innovation in a museum of this kind. The re-introduction of human facilitators, and the discreet nature of the tools used to present information, will help safeguard the fantasy aspect that visitors to the Musée des Arts et Métiers aspire to, while providing them with the information they expect.

KNOWLEDGE NETWORKS

Whatever the museum's ambition may be, it can and must not act as a substitute for other links in the chain of knowledge: schools, media, and the professional environment.

The Musée des Arts et Métiers is at the heart of a network of museums and technical cultural centers, and has chosen to make reference in its permanent exhibitions to professions, training, companies, and research laboratories relating to each of the seven main sectors. The *kiosks* are evidence of this desire for openness, which also expresses the need for an active partnership with the world of industry and research. The "kiosk" is a clearly identified space, easy for visitors to recognize. Located in the fourth—or contemporary—period of each sector, it offers an opening onto the working world, places of interest, and collections presented in other museums. In the "Mechanics" sector, for example, one item is proposed for each major family of professions; "Scientific Instruments" presents a journey through reference museums: Florence, Leyde, Kassel, Oxford and Cambridge; and the "Building" space is structured around the major architectural projects recently carried out in France. The "kiosk" also refers to the Documentation Center for the general public, which contains material useful for research or information, complementing the trails followed through the permanent exhibition.

The program of *temporary exhibitions* also helps to nourish exchange networks between the Museum and its environment. It will be a source of enrichment in the future, thanks to a reasoned and deliberate policy of acquisition and conservation of the key elements in our contemporary technical heritage. In the choices they make, the curators require extensive advice and guidelines, since the field to be covered is enormous. In what way, for example, should we conserve software, without which the machines making up increasingly dense computer networks are deprived of their "soul"? The temporary exhibitions area is located close to the main reception and can be accessed separately. These exhibitions are usually co-produced with outside partners, mainly companies and other art and technical museums. They provide an opportunity to present some of the pieces kept in the reserve store, and are an incentive to acquire instruments or machines suitable for integration into the Museum's collection.

Located within a large teaching and research establishment (the CNAM currently trains almost 100,000 people per year), the Musée des Arts et Métiers is at the very heart of the educational system. While it can only welcome a limited percentage of any given age group in quantitative terms, its role can be exemplary in promoting a method of learning based around a sensitive and curious approach to technical objects. *Educational workshops*, located next to the exhibition rooms, will allow small groups of adults or children to make their own machines or demonstration models. Giving people a taste for experimenting, handling, and creating is the aim of the workshops, and they can function either as a complement to a visit or in a more independent capacity. A series of "*Carnets pédagogiques*" ("Educational Notebooks"), published by the Museum, are already preparing the way for the implementation of a workshop-type resource, re-situating objects within their context while explaining the possible ways in which they can be made ("Foucault's Pendulum," for example.)

A place of *memory and learning*, the Museum is also a place of exchange, thanks to the development of information networks such as the Internet on a global scale. These "screens of knowledge" are not a rival to the Museum. On the contrary, they help in promoting and enhancing the institution, thanks to its image banks, its collection of objects and drawings, its documentation, and its archives. These virtual museums create a richer vision of the real museum which, if it is to last, must be rooted in its site, in its memory, in its displays. During the renovation work, the Musée des Arts et Métiers chose to develop an ambitious editorial policy, in particular through the creation of a quarterly *Revue*, devoted to the history and museology of technology, a CD-ROM presenting the rich variety of the collections, and a journey through the history of technology (*L'Album*), and by proposing a visit of the Museum on the Internet, giving people access to its entire "object base" of over 80,000 pieces. This virtual and interactive visit paves the way for the discovery of new exhibition spaces in 1998. These will necessarily be more selective and the conservation team in charge of the renovation program must measure and assume responsibility for the choices made.

"Libraries and museums formed selectively are in a sense the workshops of the human mind" (Henri Grégoire, 1794). As a workshop-museum, the renovated Musée des Arts et Métiers wishes to remain as close as possible to people and their creativity. In this respect, it remains loyal to its original vocation: to demonstrate the treasures of ingenuity buried in the heart of mechanisms, within machines.

While these "mechanisms" have now become immaterial, an "internal" vision is still required. The software industry is creating new technological frontiers, new relationships between man and machine. Far from losing its *raison d'être*, the Museum can and must remain a source of inspiration and a place for comparing new technologies. "All the arts have points of contact," wrote Abbé Henri Grégoire, two hundred years ago. As a global museum, the Musée des Arts et Métiers, now a bicentennial establishment, is directing its renovation toward becoming a place where history meets the contemporary world, where art meets science, where memory meets imagination.

NOTES

1. See the study by Lucien Mironer, Aymard de Mengin and Agnès Suillerot, *Fréquentation, notoriété et attraction de sept établissements culturels parisiens* (Paris: Cité des Sciences et de l'Industrie, Mission Musées, 1995).
2. *Arts et Métiers, the Museum Album*, CD-ROM Mac-PC, Musée des Arts et Métiers, Productions La Forêt, 1995.
3. "An interview with François Dagognet," *Le Monde*, November 2, 1993, p. 2.
4. *Les Réserves dans les Musées, The Reserve Collection in Museums*, international symposium, September 19–20, 1994 (Paris: Musée National des Techniques, 1994).
5. Pierre Piganiol, *Le Musée du Conservatoire national des arts et métiers, sa renaissance, pourquoi? comment?* (Paris, 1989).
6. See Hana Gottesdiener and Jean Davallon, "L'Homme Machine, an exhibition, its publics," *Musée des Arts et Métiers, La Revue*, no. 11 (June 1995): 13–19.

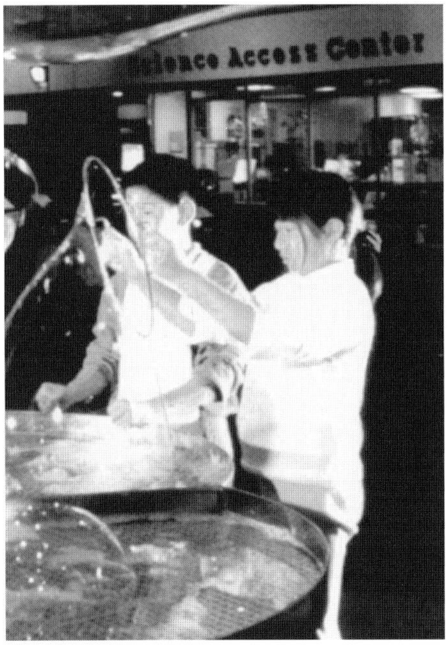

Museums, Communities, and Contemporary Science

Alan J. Friedman

Science museums are a billion-dollar-a-year industry, supported by a mixture of communities: government, corporate, individual, and foundation, as well as the general public. Science and technology centers are a relatively recent subset of the science museum realm, with a specialization on "hands-on" or "interactive" exhibits, and a spectacular record of growth. In 1972 there were 17 institutions that identified themselves as belonging to this category. Today there are 450 such science-technology centers in the world, most of them created within the past 20 years. In the United States alone they draw over 100 million visitors a year.[1]

Along with the generally happy situation, however, there have been some serious questions about the roles and effectiveness of science museums and science centers, and a few disturbing near-failures. A dozen science museums have found themselves in life-threatening situations, most often but not always beginning with overruns in the costs of initial construction or of a capital expansion. These examples of serious problems, facing such well-established museums as the Franklin Institute, the Oregon Museum of Science and Industry, the Smithsonian, and highly promising newcomers like the Columbus Center, Liberty Science Center, and newMetropolis, suggest that it is important not to take community support for science centers and museums for granted. Indeed, some unexpected event, like a cost overrun or a controversial exhibition, may cause the various communities of stakeholders to discover suddenly that they each have wanted different things from these institutions.

This paper discusses four major communities that are critical to science museums, and offers views of where the interests of these communities coincide and where they differ. These four constituencies are scientists, public visitors, funders, and museum staff. This analysis is based primarily on the author's experiences at his own institution and at other U.S. science centers and science museums. The term "science museum" is used here to refer to all types of science-technology museums and centers.

WHAT SCIENTISTS WANT A SCIENCE MUSEUM TO DO

Scientists and scholars want to educate the public about the fundamentals of their fields. They want museums to explain the processes through which science creates and learns. They want to celebrate the achievements, both historical and contemporary, of which they are most proud.

Scientists and curators also want contemporary science content. The major events of 20th-century science, including relativity, quantum theory, DNA, and the expanding universe, are among the topics that scientists believe should be covered.

WHAT VISITORS WANT A SCIENCE MUSEUM TO DO

The public that visits museums is not, in general, looking to learn any specific information. Family visitors say they want a pleasant, entertaining afternoon together. Individual communities want to see celebrations of science and technology related to their own culture and history. There is one overriding demand: to explain how the subject directly relates to the visitor's own life, and what the visitor already knows and is interested in.

When the New York Hall of Science did extensive "front end" evaluation for an exhibition on microbial life, and another on chemistry, we learned that visitors were almost exclusively interested in those microbes that they had heard about in the mass media recently, such as the viruses which caused AIDS, or which had affected them directly, such as the common cold. Visitors were also interested in the chemistry of their own bodies. They did not volunteer a strong desire to learn about the underlying principles, the classifications and terminology, or the processes of investigation that scientists used to learn about microbes.

WHAT FUNDERS WANT A SCIENCE MUSEUM TO DO

In general, funders want to be associated with excellence and success. An institution with strong visitorship, an excellent public reputation, and a healthy financial situation is likely to attract government, business, and individual support. The funders want their own judgment in recognizing excellence, and their generosity, to be recognized and celebrated as well.

Beyond these general attributes, however, funders are also looking for institutions to contribute to the quality of life of their communities in several specific ways. The government of New York City, for example, is most impressed by the role museums play in stimulating the City's economy. The Cultural Institutions Group of New York City, which represents 34 museums receiving the most funding from the City, estimates that the US$ 84 million that City government spends on these museums plays a key role in generating US$ 2,000 million in economic activity. This activity includes the direct and indirect impact of spending by the institutions themselves, and increased tourism.

Quality cultural institutions also contribute to the residents' sense of well-being in their neighborhoods, making people happier to live in a particular region. This argument is appreciated by corporations that want to attract and retain employees, and it is appreciated by businesses such as real estate investors.

WHAT STAFF WANT A SCIENCE MUSEUM TO DO

The museum exhibits and education staff of science museums are acutely if quietly aware of the limitations of the exhibition medium. First, many exhibit developers hold that what counts is not what was *intended* by a design, or even what an exhibition *presents*, but only *what the visitor actually takes away from the experience.* Science-technology museum visits typically last one to three hours; individual exhibit units typically hold visitors for between 30 seconds and 3 minutes. Within these constraints, exhibits can make limited (if lasting) impacts on visitors' knowledge and attitudes.[2]

Exhibits can be expected to communicate only simple stories. "Hands-on" exhibits are best at presenting the sensory impact of real objects and phenomena. You can see, hear, and smell the spark from a Van de Graff generator; feel the effect of a compound pulley; watch in real time the unpredictable life and death of individual microbes. Simply putting a phenomenon or object on display celebrates it, but usually fails to be effective in placing a phenomenon in a broader or deeper context. Exhibitions that try to do too much may only confuse or intimidate most visitors, who enter a science museum somewhat timidly anyway.

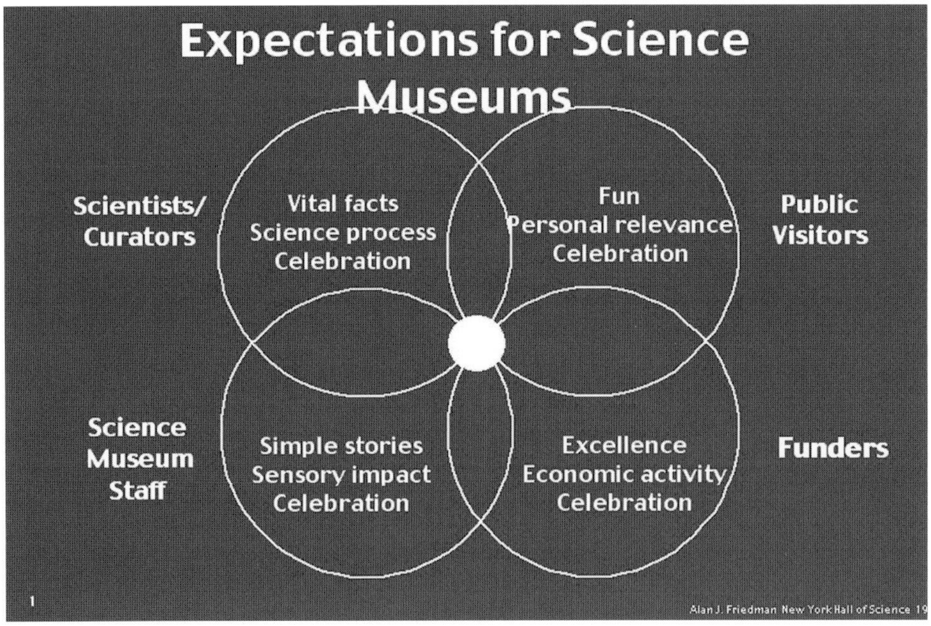

CHART 1.

WHERE THE EXPECTATIONS OF THESE FOUR COMMUNITIES OVERLAP

These four sets of expectations overlap primarily in the word "celebrate" (**Chart 1**). When they find the right things to celebrate, science museums have a chance of pleasing everybody a little bit, even if they reach only a small part of the full potential each of the stakeholders wants to achieve. Most science museum exhibits today are, indeed, celebrations of some object or phenomenon. Celebratory exhibits normally present little controversy, but these exhibits also usually present little contemporary science, little opportunity to understand the processes of science, little of the broader context that may be required to make the subject of an exhibit truly important.

COMPONENTS OF EXHIBITIONS THAT PLEASE EVERYONE

Gyroscope (Figure 1, at Techniquest in Wales): The bicycle wheel gyroscope celebrates a very

FIGURE 1. Bike Gyro exhibit at Techniquest, Wales.
Photo: Alan J. Friedman

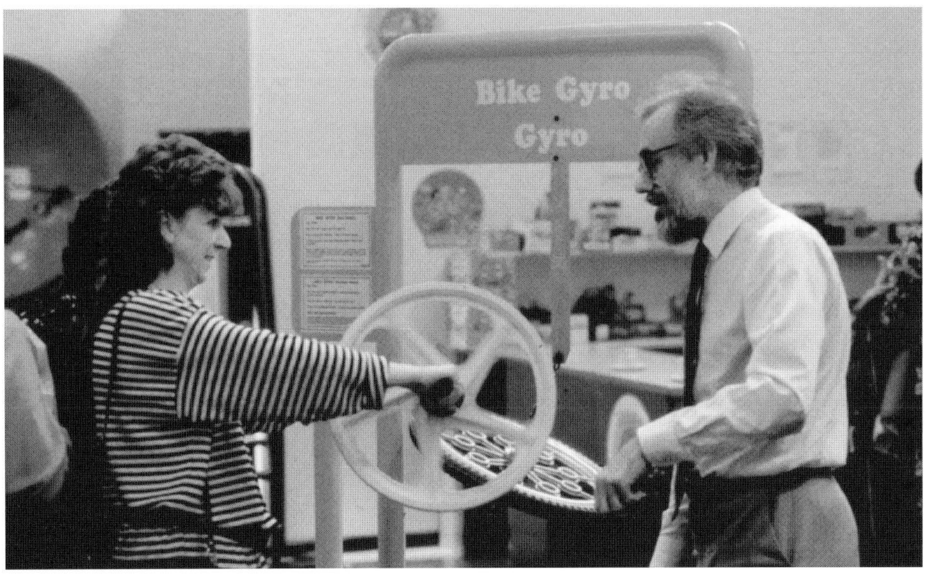

fundamental phenomenon, and because it behaves in an unexpected manner, the device also delights visitors. Industrial applications of the gyroscope are part of several major achievements in contemporary technology.

Soap Bubbles (Figure 2, at the New York Hall of Science): Making giant bubbles illustrates several important phenomena, although few visitors know the language scientists use to describe those phenomena. Nevertheless, the popularity of bubble phenomena is strong among the public because the sense of accomplishment, esthetics, and connections to everyday experiences make the bubbles accessible to everyone.

Catenary Arch (Figure 3, at the Exploratorium in San Francisco): Like the gyroscope, arch bridge exhibits, both large and small, demonstrate something important in science and mathematics,

FIGURE 2. Soap Bubbles exhibit at the New York Hall of Science, part of the *Seeing the Light* exhibition created by the Exploratorium, San Francisco.

Photo: Ken Howard

FIGURE 3. Catenary Arch exhibit at the Exploratorium, San Francisco.

Photo: Alan J. Friedman

with successful applications in industry and in everyday life.

EXHIBITIONS THAT FAIL TO MEET THE EXPECTATIONS OF ONE OR MORE COMMUNITIES

Quantum Atom Model (Figure 4, at the New York Hall of Science): A three-dimensional, interactive, dynamic model of quantum energy states of a hydrogen atom is very pleasing to scientists and serious students, but has proven to be difficult for the general public to appreciate or interpret. Too much prerequisite knowledge is required to understand why quantum theory was so surprising to scientists early in the century, and why the behavior it depicts is so radical and important to our understanding of the universe. The major intellectual achievement this concept represents is largely unknown to visitors. Visitors lacking some prior awareness of the revolutions in twentieth-century physics have a difficult time understanding what all the fuss is about.

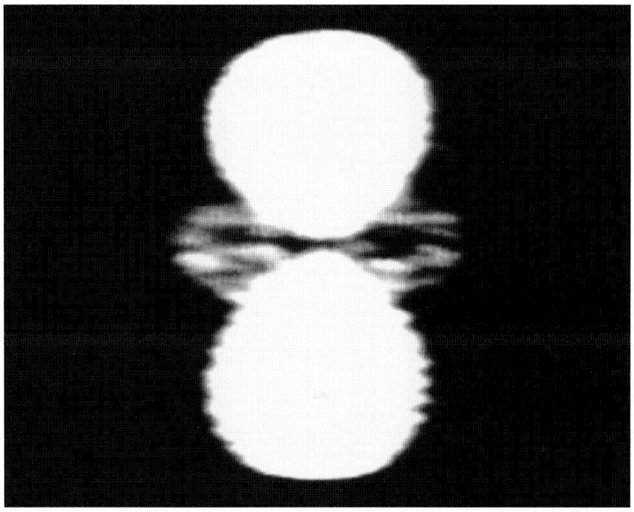

FIGURE 4. Quantum Atom Model, part of the *Realm of the Atom* exhibition at the New York Hall of Science.
Photo: Alan J. Friedman

Einstein (Figure 5, at the Exploratorium in San Francisco): Despite Einstein's fame and personal attractiveness to the general public, it has proven extraordinarily difficult to explain what he did to non-scientists. The science center most accomplished in presenting real phenomena in accessible form, the Exploratorium, was able to produce very little in the way of demonstrating Einstein's work itself. Its Einstein exhibition primarily presented phenomena that exposed inconsistencies in 19th-century theories. Most visitors, however, were equally unfamiliar with either 19th-century or 20th-century physics. Like the quantum atom model,

FIGURE 5. "Thermionic Emission," part of the *Einstein in Context* exhibition at the Exploratorium, San Francisco.
Photo: Alan J. Friedman

this exhibition is pleasing to scientists but of little interest to most of the public.

Enola Gay (Figure 6, at the Smithsonian's Air and Space Museum): The *Enola Gay*, the airplane that dropped the first atomic bomb on Hiroshima, is a concrete object representing a pivotal event in 20th-century history and in the entire history of the relations between science and society. The Smithsonian Air and Space Museum's attempt to place this object in a cultural and historic context led to threats to cut off government funding and eventually resulted in the resignation of the museum's director and the presentation of the airplane in a practically "interpretation-free" exhibition. While there were real intellectual disagreements about the best interpretation of the Hiroshima bomb, the failure of this planned exhibition was largely due to its failure to fit within the "celebration" expectation that funders and some sectors of the public audience regarded as the primary mission of this museum.[3]

FIGURE 6. Enola Gay aircraft, part of the *Enola Gay* exhibition at the Smithsonian National Air and Space Museum, Washington, DC.
Photo: Alan J. Friedman

Science in American Life (Figure 7, at the Smithsonian's Museum of American History): This exhibition, like the planned *Enola Gay* exhibit, had similar problems. The curators had a particular view of one aspect of science in American life, a view that stressed the public's increasing disenchantment with science and technology as "cure-alls" for social problems. The major non-government funder of the exhibit, the American Chemical Society, had expected a celebration of the accomplishments that science had in fact achieved during the history of the United States, and were disappointed with the preponderance of negative examples that the curators used to make their more tightly focused interpretation clear. Members of the American Physical Society, a group of scientists who had not been funders, were even more outspoken in their criticism, vigorously articulating the view that the function of museum exhibitions is solely to celebrate accomplishments. While visitors to the museum were overwhelmingly pleased with the exhibition, rarely noticing its underlying theme, the unhappiness of scientists and scientist/funders created significant commotion and threats to the museum's funding.[4]

COMMON CHARACTERISTICS OF "SAFE" EXHIBITIONS

Success or disappointment with any given exhibition may certainly be related to the particular execution of that project, as well as its more generic properties. Such limitations notwithstanding, most successful universally pleasing science and technology center style exhibits have been essentially *non-verbal, non-sequential* experiences for visitors. In this fashion, they are more like art museums. Works of art themselves are largely meant to be non-verbal, non-sequential; so art museums can function merely by placing objects on walls, with little or no signage or context. The same thing can be done with the phenomenological exhibits at science-technology centers.

But science, especially contemporary advances in science of the type that win Nobel prizes, and understanding of cultural or historic context, are highly verbal (the language is sometimes mathematics) and require prerequisite, sequential knowledge to be understood. Objects like a computer chip, a telephone switching station, a recombinant

FIGURE 7. Laboratory Science Comes to America/1876–1920, a section of the *Science in American Life* exhibition, Smithsonian National Museum of American History, Washington, DC. *Photo: Alan J. Friedman*

DNA laboratory, or a theoretician's brain do not "speak for themselves." So science-technology museums, and science centers in particular, have a very difficult time presenting contemporary science, the process by which science is done, or ideas about social context.

SUGGESTIONS FOR MOVING BEYOND CELEBRATION

There are many topics and goals for exhibits sought by scientists, scholars, funders, and the public that are extremely difficult if not impossible to communicate through the exhibit medium, and perhaps are just better suited to books, television programs, or formal classrooms. Nevertheless, science museums, to meet their own ambitions and those of their constituent communities, are attempting to move beyond celebration. One strategy to handle difficult topics and goals, including prerequisite knowledge and sequence, is to create simulations or even theme park rides, to impose an order to the visitor's experiences. But these presentations usually have low "reality quotients": visitors learn mostly about the rules of the simulation, not the rules of nature. These exhibitions may also break faith with the visitor's expectation that, in an informal learning environment, it is the visitor who should control the order and timing of the experience in the museum.[5]

There are, however, some possible solutions available that use science-technology museum strengths and still communicate effectively verbal, sequential stories:

- the use of multiple media so that the physical, three-dimensional objects in an exhibition and the labels do not have to carry the sometimes impossible burden of satisfying conflicting expectations. Random access and layered audio tours (**Figure 8**, New York Hall of Science), live tours, theatrical presentations on the exhibit floor; all allow more complex, multiple verbal perspectives to be offered and multiple agendas to be addressed, and early controlled experiments suggest these techniques can be effective.
- the creation of guided, longer duration, but still free-choice experiences, such as the experimentation galleries at the Boston Museum of Science and the Science Museum of Minnesota.
- exhibitions that make use of the social interactions among visitors and among visitors and staff, to explore contexts and controversial issues. "Mine Games" at Science World in Vancouver **(Figure 9)** has demonstrated the possibility of effectively dealing with conflicting interpretations

FIGURE 8. Visitors in the Seeing the Light exhibition at the New York Hall of Science, using a random-access audio tour during an evaluation study.

Photo: Teri Bloom

FIGURE 9. Section of the Mine Games exhibition at Science World, Vancouver.

Photo supplied by and used with the permission of Science World, British Columbia.

and biases in contemporary issues where science, technology, and other communities of interest meet.

Science museums are learning, albeit slowly and with some pain, how to combine the pleasures and successes of science-technology museum exhibitions with the inherently verbal, sequential structures required to go beyond the desirable but limited goals of celebration of science and of nature.

NOTES

A shorter treatment of this theme was given by the author for a panel at the 1998 meeting of the American Association for the Advancement of Science.

1. *Yearbook of Science-Center Statistics* (Washington: Association of Science Technology Centers, 1997), p. 19.
2. Valerie Crane, Heather Nicholson, Milton Chen and Stephen Bitgood, *Informal Science Learning* (Dedham, MA: Research Communications, 1994), pp. 61–106.
3. Mike Wallace "The Battle of the Enola Gay," *Museum News* (July–August 1995): 40–45, 60–62. See also Martin Harwit, *An Exhibit Denied: Lobbying the History of Enola Gay* (New York: Copernicus Books, 1996).
4. Jane Gregory and Steve Miller, *Science in Public* (New York: Plenum Press, 1998), pp. 214–219. See also Robert L. Park and Ursula Goodenough, "The Unmaking of American Science Policy: The End of the Scientific Era?" *Academe*, (January–February 1996): 15; Andrew J. Pekarik, Zahava D. Doering, and Adam Bickford, "An Assessment of the 'Science in American Life' Exhibition at the National Museum of American History," Institutional Studies Office, Smithsonian Institution, Report 95-5 (November 1995); Joan E. Shields, "Science in American Life Revisited," *Chemical and Engineering News*, (March 11, 1996): 40; Alan J. Friedman, "Exhibits and Expectations," *Public Understanding of Science*, Vol. 4 (1995): 305–313.
5. Alan J. Friedman, "Differentiating Science-Technology Centers from Other Leisure-time Enterprises," *ASTC Newsletter*, 24, No. 1 (1966): 7–10.

The Level of Complexity: The Middle Way Between the Superficial and the Repellent

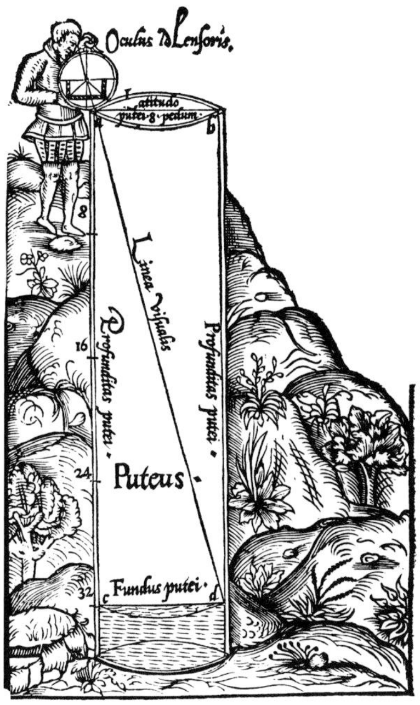

Beyond Understanding

CURATORSHIP AND ACCESS IN SCIENCE MUSEUMS

Jim Bennett

A museum of modern science presents a range of difficult challenges to its curators. Some are associated with the objects themselves, such as their size and safety, or their uninformative and unpleasing appearance. Others are administrative and financial, with cost commitments far outstripping any prospect of revenue. Still others concern public indifference, bafflement or suspicion. But one particular difficulty that curators will cite is public access, not just to the objects but to what they will call the "science." Science is difficult, obscure, complex, abstract and counter-intuitive. It seems all but impossible to take the lay visitor into a meaningful understanding of modern science. The idea of finding a "middle way" takes this as the central dilemma of science curatorship: very simple explanations are useless distortions, full explanations are impossible.

But is it really correct to think that our situation is any more challenging than that of many other museum curators? Consider the prospect of mounting an exhibition on state bureaucracy, contemporary warfare, newspaper production, transport systems, library management, and so on. How might we set about making a gallery presentation of a foreign language or a great opera house? There are many complex systems in the modern world, and many specialist understandings to cope with them, but it is in relation to science that in recent years there have been particular anxieties about "public understanding," and it has been in museums, or quasi-museums, that these anxieties have had their most visible effect. Museums of *modern* science are particularly implicated in this development, since the public understanding program is interested only in contemporary understanding. In fact, the characterization should be restricted further to contemporary understanding within a particular cultural context; the understanding being offered is bounded in time and in culture.

Everyone can agree that public understanding of science is a good thing. Public understanding of most things is a good thing, and museum professionals are in the business of promoting understanding for themselves, for their academic disciplines, for their specialist interest groups and for the general public. The problem is rather whether we can agree on the nature and character of this "understanding." The move from an unexceptional platitude to a program of action is too easily taken for granted and it is important not to allow particular interests and agendas to slip by uncritically; in this case these are embedded in assumptions about what constitutes "understanding."

One way of approaching the issue is to recall the influential remark of one of the founders of the science-center movement in Britain, Richard

Gregory, who declared that traditional science museums contained "remarkably little science."[1] At first sight, it is not clear what Gregory and other critics who followed this line could have meant. The museums and exhibitions they were criticizing were full of the tools and products of science, since these traditional museums tended to have rich displays of instruments and apparatus used in scientific experiment, in demonstration and teaching, and in the professional and industrial applications of scientific techniques. Was this not a display of "science"?

The critics felt they had two justifications for answering "no"—one temporal, the other conceptual. Part of the problem they saw was that these displays were overly focused on the past. But, more importantly perhaps, it was thought that there was something behind all this materiality, something structural, abstract and conceptual, something whose intellectual integrity was independent of these material relics and applications, and that something was the "science." The science could stand independent of such relics; understanding it would comprise, not a familiarity with things, but a grasp of concepts and their relationships. Even if this is a caricature that applies only to the more extreme positions adopted by these critics, it does represent the shift in emphasis and direction demanded and subsequently effected by the public understanding program.

To some extent, the approach to exhibition building offered as an alternative was advanced at the expense of the traditional museum values of the care and display of collections. There was, and sometimes there still is, pejorative talk of lifeless displays, silent objects, closed showcases, confined visitors, artifact fetishism, and so on. Interactives and "hands-on" exhibits were lively, engaging, even democratizing; more traditional displays were dry, austere and alienating.

But objections to presentation and display technique masked a more fundamental move to close down the range and scope of worthwhile "understanding." There are many ways of engaging with and entering into scientific culture, and the scientific relics in museum collections and displays raise many different kinds of questions about science alongside those concerned with how scientists codify and express their beliefs about the causal structures of the natural world. For a single-minded mission to explain "the science," objects are problematic because of their ambiguity and the richness of their associations for the viewer: their meaning and significance are not fixed, and visitors' reactions to them are difficult to control.

As nothing more than a parable, to illustrate the idea of closing down the range of understanding, we might take an example from one of the other subjects concerning difficult displays mentioned above. Imagine a visitor to an exhibition related to, say, European railways, a visitor with little or no relevant experience. Now imagine that the exhibition we have designed for this visitor presents a set of hierarchical, schematic maps of the railway network, and sets of outline timetables to show the basic structure of how the rolling stock is managed, perhaps with sub-sets of national and regional maps and schedules, offering more detailed case-history explorations of particular areas. Imagine that we have been clever enough to let the visitor try out a computer simulation whereby certain network configurations and timetable adjustments can be chosen and their consequences played out. Given sufficient application, it would be possible to emerge from such an exhibition with a fair grasp of at least the basic principles of such a system and a familiarity with some of its applications. No-one for a moment would judge that more than a very impoverished "understanding" of European railways would have been acquired. We would want our visitor to know something about trains, their locomotives and carriages, their manufacture, about the logistics of fuelling and maintaining the trains, the companies who run them, the government policies that legislate for them, the experience of passenger travel and its social implications, the economic significance of freight services, the history of network expansion and contraction, and so on.

As an alternative thought experiment, we might consider mounting a "science" exhibition about the distant past. Is it really a simpler task to introduce the visitor to, say, the medieval astrolabe? Compared with recent material, the technical content may well be very much more elementary, even though it still seems to baffle almost everyone, but many other interpretative problems have been magnified across such a distance in time. The disciplinary context for the instrument's use is quite unfamiliar, the social context for its commissioning, manufacture, retailing and possession are completely foreign, the functionality of the instrument is not what we are inclined to assume, the iconography is exotic, and so on. The interpretative advantage we may have gained in one respect has been at least offset by disadvantages in others, unless, of course, we restrict the "understanding" being offered to conceptual and technical matters, unless we value only a command of "the science" in the narrow definition encountered above.

The restricted notion of understanding that has been applied to many science displays allows the exhibition agenda to be carefully controlled. Exhibitions in every field adopt positions and postures, but outside science there is much more tolerance of visitors' own agendas and greater equanimity about visitors leaving unconvinced by or even hostile to the curatorial account of what they have seen. A sceptical visitor to an art exhibition is completely accommodated by the expectations of the occasion, on the part of the curator as well as herself. But a visitor can hardly leave a science exhibition saying that she did not "take to" the science on display and would have preferred a different one. Visitors know, of course, that this would not be a proper response, so instead they frequently *apologize* that they have not "*understood*."

In a sense, and to a limited extent, we expect visitors to *become scientists*. We measure their success as visitors, and our success as exhibition builders, by how far they have been able to enter into the belief system characteristic of the science in question, to follow in the footsteps of those who actually make use of this system professionally. Again, we do not have similar expectations in other fields. It is of no concern to us if visitors to art galleries have made no progress as painters or sculptors. We would probably be surprised if visitors to anthropology galleries were moved to belief in an exotic religion whose liturgical artifacts were on display.

As for "democratization," we should distinguish between the packaging of exhibitions—the presentation and display technology—and their content. Certainly an inviting and engaging display will have the desirable effect of making visitors feel that the museum has something for them, and that, of course, is what we should be aiming at. But we surely cannot deceive ourselves into characterizing this as a democratic involvement with science. The educative experience offered to the visitor—the *content* of the display—is directional: it proceeds so far as is possible along a determined path. By setting the criteria of success, by prescribing a single agenda, by circumscribing the meaning of "understanding," we limit the possibilities for the visitor; we close off different responses to the collections held by the museum. This may be just what is needed for making people scientists of a sort, but we should not delude ourselves into thinking that it makes science more democratic.

Should we, on the other hand, think it strange that this closing down of alternative agendas has taken museums of science in a direction away from the museological concerns of professional curators in other fields? The debate elsewhere has involved the recognition that collections and displays have interests that can be located in history and in culture, that objects have meanings that are differentiated across contexts both outside and inside the museum, that exhibition agendas may well be unavoidable but can be recognized and conceded, and need not obligate the visitors, who can adopt alternatives.

Museological discussion in science has instead been more concerned with effectiveness. The new display philosophies were criticized, certainly, but at first on the grounds that they did not deliver what was promised. Having fun was all very well, but it no more delivered an understanding of "the science" than the static presentation of collections had done. It seemed to be taken for granted that there was an agreed end, and debate only over the means.[2]

Thankfully the museological debate in science is gathering strength and sophistication, and a series of recent anthologies has shown that commentators on museums and exhibitions in our field are prepared to take on more fundamental questions than how successful they have been at achieving their aims.[3] The most recent is a set of essays published in French under the promising title *La révolution de la muséologie des sciences*. Here James Bradburne, for example, who has been closely involved with the development of science centers, particularly and most recently with "newMetropolis" in Amsterdam, enumerates a set of criticisms of such institutions: that they communicate the principles and not the processes of science, they are "textes de physique en trois dimensions," they neglect the links between science and technology, and so on, and in general that the impression of exploration and discovery is bogus, since the underlying exhibition agenda is not negotiable: "Le message sous-jacent est que seuls les scientifiques sont en mesure de définir les objets de la science."[4]

Bradburne speaks of a "crisis" in science centers, one precipitated by falling visitor numbers. He believes that something new is needed, and he offers a critical analysis in which we find that there is already a category for something he calls the "traditional" science center—one where individuals confront isolated interactives in an arrangement that does not encourage discussion between visitors, one that insists on pure science even though visitors are more interested in its social aspects, and one that offers fun as a means of overcoming established antipathy to school science, but which at the same time is based on a similar schoolmasterly attitude to learning: "tout apprentissage est de type coercitif, du haut vers le bas." His summary characterization of the traditional science center is "un malheureux mariage de stratégies pédagogiques scolaires dans un environnement d'apprentissage non scolaire."[5]

The imminence of a new century seems to have encouraged other commentators in this volume to look forward to future institutions, to what might replace not just the traditional science museum, but the traditional science center. Emlyn Koster, for example, presents an account of science museums in three successive generations. The first is based on collections and the visitor's role is "regarder silencieusement des vitrines." The second—the present generation—is characterized by participation, movement, gadgetry and interactives. The third, it turns out, is more difficult to define and the characterization is nebulous, comprising pious aspirations rather than practical programs for realizing them. Whatever it is, people will be more important, society will be brought back into the scientific enterprise, science and culture will interact, and scientists and citizens will come together to face the technological and environmental problems of the twenty-first century. And whatever it is, some institution evolving from the science center will be at the heart of this new realization of a scientific culture.[6]

Museum taxonomies, developing over time, are offered by a number of contributors, with "generations" adopted as a common metaphor. Those identified by Bernard Schiele, for example, also number three, again with the science center as the second generation, giving way in its turn to a third, which again is a pious invocation of indisputably worthy things, such as wholism and integration.[7]

If we return to Bradburne, he too has ambitious expectations of the next generation of public institutions. In connection with the hopes for newMetropolis, we learn that it is tailored to a reformed society, not one comprised of producers and consumers, but "une société de personnes qui travaillent et qui choisissent." In such a society, the science center must be seen "non comme un sous-produit, mais comme un intermédiaire indépendant, qui favorise une pensée critique, curieuse et créatrice."[8] Bradburne says that museums and science centers have for too long operated in the shadow of the preoccupations of other groups, and that in the future they can have an independent role. But once again this useful idea is not cashed out beyond what can be read, to speak frankly, as a set of unexceptional platitudes.

Quoting from the founding policy of newMetropolis, Bradburne gives us some stirring ambitions for the future:

> De plus en plus, le centre scientifique offrira un cadre social que l'on ne trouve pas au foyer. Le centre scientifique sera un endroit de rencontre, de discussion, d'échange d'idées et d'information. Ce sera un lieu ouvert à tous, un forum, une place publique. Connecté aux réseaux d'informations internationaux, le centre scientifique deviendra un carrefour social et culturel, où les gens peuvent communiquer—seuls et en groupe—avec d'autres personnes de la région, du pays et des quatre coins du monde. C'est cette dimension sociale, dans son sens le plus large, qui distinguera le centre scientifique des autres possibilités qu'offrent la ville ou le foyer.[9]

How, in any practical terms, all this might be achieved by the third generation of science museums is far from clear; how the agenda of the science center is to be modified to take on this extraordinarily ambitious social program is not explained.

So, one of the strong impressions from this anthology concerns the need for the science center in making traditional and elegant glass souvenirs for its visitors, and the gorgeous green-glass torus then made for Hannes Alfvén. The natural magic of the most up-to-date science centers often trades on the Wagnerian imagery of workshop and cabinet proper to this allegedly long-gone milieu. Around 1900 laboratories and lecture theaters were stocked with sequences of devices akin to traditional objects of brass and glass. There were rather few firm distinctions even between instruments of precision measurement, those of didactic demonstration, and the objects proper to natural magic and the cabinet. Objects used for teaching scientific principles could be as sophisticated and complex as those of backstage research.[16]

As Paolo Brenni has pointed out, the mid-twentieth century gradually saw such objects made of plastic rather than metal, stripped of ornament and much more firmly encased. The rapid growth of the electronics industries, especially the high-performance vacuum tubes used in radio and lighting systems and well-charted in Lindqvist's stories of Geissler tubes, were both fundamental for the new physics and, following the miniaturization and automation offered by semiconductors, led to marked changes of scale and scope in the hardware of the laboratory. Frank Oppenheimer long resisted the introduction of electronics and computers into the Exploratorium. "The uninformed observer of an electronics laboratory sees only closed boxes with gauges marked by trembling index needles and hears only unintelligible sounds. Laboratory workers sit silently, intent on oscilloscope screens. Few outsiders can identify the workers' movements or participate in their intellectual activity." Faced with the growing significance and prestige of Silicon Valley, however, the Exploratorium began incorporating "an enclosed electronics lab" as a permanent installation from 1981.[17] The development can well be compared with recent museum-based enterprises, in Manchester Museum of Science and Technology as well as the Science Museum, for example, of rebuilding early computers. As Jon Agar has noted, while such computers are "archetypical modern objects," reconstructions of early computing machines already summon up a nostalgic spirit of prototypical exploration and innovation in the name of precise replication of such devices. Agar conjectures that science museums can here "play a crucial domesticating role." The carefully-defined gap between current computer research and the show of working replicas preserves a sense of history and culture without threatening the equally important image of constant technological advance.[18]

The postwar development of digital processing and graphic display completed a process in which the furniture of the laboratory became complex arrays of apparently interchangeable signal devices and detectors. Here is how a journalist described a central workroom at CERN in 1984: "It is located on the ground floor of a two-story building that is simply two corridors one atop the other . . . Inside are scattered half a dozen computer terminals. In the center of the room sit two large pieces of technology that look as though they were made for the ultimate video game; these are Megateks. The Megatek is a computer system that displays the collisions of protons and antiprotons in three dimensions." Megatek Graphics is in fact a division of Space Electronics Inc., an integrated circuit manufacturer that produces a line of radiation-hardened microelectronics for use in commercial and military satellites. In a manner made familiar by Peter Galison's remarkable recent account of computer simulations and automation in high-energy physics, this kind of object was part of "a history of anxiety." Control, discovery and the objects themselves all seemed during the later decades of the century to be moving away from a classical culture of homely and, as Galison rightly urges, mythologized scientific workshops. It was claimed at CERN that "Megateks were the very same devices that science-fiction film makers used to produce their space battles in films like *Star Wars*"; it was also pointed out by hostile bubble-chamber physicists at CERN that "too much weight has been given to the Megatek as an analysis tool." Such hardware, and such conflicts, provide the stuff of the large-scale scientific culture which has influentially been understood as a "market-conscious, product-oriented, and capital intensive activity, which has taken on the impersonal and inhumane nature of industrial enterprise."[19]

So public science holds that in the modern age it has become too complex, esoteric and alien to be understood; and that exhibition of its mundane material culture is not the right way to mobilize understanding or interest. The force and oddity of this problem can well be illuminated by a comparison with the predicament of museum ethnography. Shows in anthropology museums routinely propose the integration of displays of artifacts with the communication of alien and unfamiliar world-systems and beliefs. Indeed, the renewed energy of debate, controversy and study in the mileu of museological anthropology reminds us at once that there is of course no necessary contradiction between exhibiting a system of knowledge judged alien and displaying a system of objects judged aesthetically appealing and dramatic. In the terms developed by Nicholas Thomas, such artifacts are "entangled objects" whose meanings shift and travel, but which therefore play a vital role in the complex web of ethnographic exchange.[20] The problem for museums of modern science seems therefore to be that one is not supposed to show the work of science anthropologically, yet it is often supposed that scientists are in the position of an alien tribe with respect to their possible publics. Anthropologists by no means possess a magic recipe which might solve the problems of science museums. On the contrary, it is precisely the obvious contrast between the situations of scientific and other communities which makes the apparent comparison of ethnographic and contemporary scientific shows so interesting. The existence of major disputes around the conservation and display of the objects of traditional cultures and those around the political meanings of recent science museum exhibitions gives graphic evidence of the relation between the two communities' problems.

Think about shows of the material culture of the sciences as elements in stories about those sciences. This is why one can correctly emphasize the need to show science in the making and the work of knowledge production through the practices and gestural knowledges which surround and give sense to objects. Such, for example, has been the thrust of the work by Otto Sibum and his colleagues on the reworking of past experiments and machines as an integral part of the exposition of scientific practice and the history of its civilizing functions.[21] Objects get their meaning partly through their integration within such stories. Many of the critical responses to anthropological and to natural scientific exhibitions develop as attacks on the stories which objects are made to tell, or which specific audiences manage to find there. The criticism then is that too many stories are being told (the exhibition is too complex); or that the wrong story is being told (the exhibition traduces the culture it purports to represent); or that the right story is not being told (the exhibition occludes what is really happening). The Smithsonian exhibition "Science in American Life" (1993) used case studies of themes such as nineteenth-century university chemistry laboratories, intelligence testing, the Manhattan Project, and biotechnology to "make science itself the central story." It got widely reported flak from established science organizations because the exhibition's story line was judged too *plural* (objects were shown in technological as well as scientific settings, for example) and yet too *selective* (for example, a member of the American Physical Society demanded that antibiotics be present to counterbalance the mushroom cloud, while other historians challenged the accuracy of the exhibition's storyline). Arthur Molella remarked that any show which thus multiplied the number of stories in which a particular scientific object was situated might now be judged a "slight to the spiritual dimension of pure science."[22]

Similar issues of the difficult multiplication of narrative contexts emerged in the Leiden exhibition of Eugene Dubois' work on *Pithecanthropus erectus*, curated there by the anthropologist Mary Bouquet the same year. In this centennial show, as the director of the Dutch Museum of Natural History put it, "scientific knowledge of human evolution is approached from a cultural perspective. Each section of the exhibition looks at the interrelationship between scientific and popular views of evolution from a different angle." The opening section, "The Movies / De Bioscoop," juxtaposed King Kong and Tarzan with the Flintstones and palaeontology. In a corridor leading to Dubois' library, the Bible lay next to the *Origin of Species*. At the entrance to Dubois' fossil collection, Bouquet put the sign banning the public from entering what had long been backstage, private research materials. Different models of ancestry, of the display of human remains, and of evolutionary descent were on show. "If New Guinean beliefs about animal ancestry could be seen as totemistic, was there not an equally totemistic element to the identification of western ancestors by western science?" It was pre-

6. The historian of physics Jürgen Teichmann has initiated several projects. See, for example, Jürgen Teichmann, "Die Rekonstruktion historischer Modelle und Experimente für den Unterricht—drei Beispiele," *Physik und Didaktik*, 4 (1979): 267–282. Jürgen Teichmann, "Das historische Experiment im Physikunterricht—wissenschaftstheoretische Betrachtungen und praktische Bedeutung," in *Wissenschaftsgeschichte und naturwissenschaftlicher Unterricht*, ed. Michael Ewers (Bad Salzdetfurth, 1978), pp. 185–212. See also note 15.
7. Falk Rieß, "Erkenntnis durch Wiederholung—eine Methode zur Geschichtsschreibung des Experiments," in *Experimental Essays—Versuche zum Experiment*, eds. Michael Heidelberger and Friedrich Steinle (Baden-Baden: Nomos Verlagsgesellschaft, 1998), pp. 157–172.
8. For the reconstructions described, see Uwe Mehrle, Jörg Meya, Falk Rieß, Gerhard Rohlfs, and Heinz Otto Sibum, eds., *From the Electrical Machine to the Theory of Ether Apparatus and Models in History of Electricity from the 18th to the 19th Century* (Oldenburg: University of Oldenburg, 1987). Among recent qualifying works in the Department of Physics, see the diploma work of Christian Sichau, *Der Joule-Thomson-Effekt: Der Versuch einer Replikation* (University of Oldenburg, 1995). See as well the diploma work of Jan Frercks, *Die Zahnradmethode zur Messung der Lichtgeschwindigkeit: Replikation von Fizeaus Experiment* (Oldenburg: University of Oldenburg, 1996).
9. See, for example, the different approaches by Stillman Drake, "Renaissance Music and Experimental Science," *Journal of the History of Ideas*, 31 (1970): 483–500; Stillman Drake, "Galileo's Experimental Confirmation of Horizontal Inertia: Unpublished Manuscripts," *ISIS*, 64 (1973): 291–305; Thomas B. Settle, "An Experiment in the History of Science," *Science*, 133, no. 1 (1961): 19–23; Thomas B. Settle, "La rete degli esperimenti Galileiani," in *Galileo e la scienza sperimentale*, eds. P. Bozzi, C. Maccagni, C. Olivieri, and T. B. Settle (Padova: Dipartimento di fisica Galileo Galilei, 1995), pp. 11–62; Svante Lindquist, *Technology on Trial: The Introduction of Steam Power Technology into Sweden, 1715–1736* (Uppsala: Almqvist & Wiksell International, 1984); Lawrence Principe, " 'Chemical Translation' and the Role of Impurities in Alchemy: Examples from Basil Valentine's Triumph-Wagen," in *Ambix*, 34 (1987): 21–30.
10. Steven Shapin and Simon Schaffer, *Leviathan and the Airpump: Hobbes, Boyle, and the Experimental Life* (Princeton: Princeton University Press, 1985), p. 16.
11. Peter Galison, *How Experiments End* (Chicago: University of Chicago Press, 1987). Compare also Simon Schaffer, "Where Experiments End: Tabletop Trials in Victorian Astronomy," in *Scientific Practice: Theories and Stories of Doing Physics*, ed. Jed. Z. Buchwald (Chicago: University of Chicago Press, 1995), pp. 257–299.
12. David Gooding, *Experiment and the Making of Meaning: Human Agency in Scientific Observation and Experiment* (Dordrecht: Kluwer, 1990).
13. On laboratory notebooks, see H. Otto Sibum, "Narrating by Numbers. Keeping an Account of Early 19th Century Laboratory Experiences," forthcoming in *Reworking the Bench*, eds. F. L. Holmes, J. Renn, and H.-J. Rheinberger.
14. The examples discussed are mainly based on my investigations of James Prescott Joule's experiment to determine the mechanical equivalent of heat. James Prescott Joule, "On the Mechanical Equivalent of Heat," in *The Scientific Papers of James Prescott Joule*, ed. The Physical Society of London (London: Taylor and Francis, 1884), pp. 298–328. For detailed accounts about the experimental performance, see H. Otto Sibum, "Reworking the Mechanical Value of Heat: Instruments of Precision and Gestures of Accuracy in Early Victorian England," *Studies in History and Philosophy of Science*, 26, no. 1 (1995): 73–106.
15. On some results of this project, see Thomas Eisenblätter and Jürgen Teichmann, *Was nützen historische Modelle und Rekonstruktionen*, Second Edition (München: Deutsches Museum, 1997).
16. We see here possibilities for co-operation with museums and academic departments. The Departments of History of Science at the University of Cambridge and Harvard University, as well as the History of Science Museum at Oxford University already work very intensively toward the integration of their instrumental resources with the practical education of the historian of science. See also two exhibitions which represent the results of such an interdisciplinary collaboration: *Empires of Physics. A Guide to the Exhibition*, eds. J. Bennett, R. Brain, K. Bycroft, S. Schaffer, H. O. Sibum, and R. Staley (Cambridge: Whipple Museum of the History of Science, 1993). *The Physics of Empire*, ed. R. Staley (Cambridge: Whipple Museum for the History of Science, 1994). As well as, *1900. The New Age*, eds. J. Bennett, R. Brain, K. Bycroft, S. Schaffer, H. O. Sibum, and R. Staley (Cambridge: Whipple Museum for the History of Science, 1994).
17. Cynthia J. Koepp, "The Alphabetic Order: Work in Diderot's Encyclopédie," in Work in France. *Repre-*

sentations, Meaning, Organization, and Practice, ed. Steven Laurence Kaplan and Cynthia J. Koepp (Ithaca, NY, London: Cornell University Press, 1986), pp. 229–257. John R. Pannabecker, "Representing Mechanical Arts in Diderot's Encyclopédie," *Technology and Culture*, 39 (1998): 33–73.

18. Michael Polanyi, *Personal Knowledge* (Chicago: The University of Chicago Press, 1958). Michael Polanyi, "Tacit Knowing," in *The Tacit Dimension*, Michael Polanyi (Gloucester, MA: Peter Smith, 1983), pp. 3–25.
19. See, for example, Mario Biagioli, "Tacit Knowledge, Courtliness and the Scientist's Body," in *Choreographing History*, ed. Susan Leigh Foster (Bloomington: Indiana University Press, 1995), pp. 69–81.
20. For criticism of a static conception of knowledge, see Ludwik Fleck, "Das Problem einer Theorie des Erkennens," in *Ludwik Fleck. Erfahrung und Tatsache. Gesammelte Aufsätze*, ed. Lothar Schäfer and Thomas Schnelle (Frankfurt a.M.: Suhrkamp, 1983), pp. 84–127.
21. See H. Otto Sibum, "Reworking the Mechanical Value of Heat: Instruments of Precision and Gestures of Accuracy in Early Victorian England," *Studies in History and Philosophy of Science*, 26, no. 1 (1995): 73–106. As well as, H. Otto Sibum, "Les Gestes de la Mesure. Joule, les Pratiques de la Brasserie et la Science," *Annales—Histoire, Science Sociales*, 4–5 (1998): 745–774.

Fact and Fancy

ART IN THE PRESENTATION OF SCIENCE

Ken Arnold

> There is no science without fancy, and no art without facts.[1]
>
> —*Vladimir Nabokov*

Why not start by confessing to a few prejudices? In time-honored fashion, however, I will endeavor to dress up what have become automatic habits of my thinking into well-thought-out theoretical presuppositions. The aim of this is to provide a platform on which to set out what I hope is a vigorous argument for the extremely rich potential within science museums to mix art in with their usual fare of science and its history.

My first prejudice can be stated as a thorough conviction that temporary exhibitions should be thought of and created in a far more experimental fashion than is common at present. If this were the case, then curators could embrace the possibilities of using exhibition projects as licenses to produce provocative essays rather than, as is too often the case at present, definitive scholarly monographs. The problem with the latter is that they commonly exhaust any but the most expert visitors and end up embodying a well-worn pejorative literary metaphor: they become books on walls. Temporary shows should instead allow curators to undertake unusual, and occasionally risky, strategies of presenting subjects and objects from striking perspectives, or in unexpected, and even quite deliberately jarring combinations.[2]

In the art world, the idea of employing exhibitions as an important agency of communication does have considerable currency. Just one particularly elegant, but rather less well known, example of what such a strategy can enable comes from Savitsky Regional Picture Gallery in Penza in Russia, where their exhibition program is based on the disarmingly simple but intriguing technique of showing just one picture at a time—unveiled behind a theatrical curtain to great effect, with a finely scripted presentation to accompany it.[3] The methodology is highly suggestive and could produce very interesting results in any number of other contexts. For very large exhibits such as boats, this methodology is sometimes adopted because no other is feasible, but even here the effects can be highly compelling. The Vasa Museum in Stockholm, for example, indicates just how powerful the full interpretation of one massive object can be. More generally, curators of art exhibitions have increasingly taken on the role of "auteurs," which in turn has posited the anticipation that their exhibitions have something idiosyncratic to say. Though this way of working has been less exploited in

science museums, there are healthy signs that things are changing there too.

My second prejudice is to claim that exhibitions, and indeed individual exhibits within them, need to be more clearly perceived as an art form rather than just a communicative strategy. In art galleries, one would never question this assumption, but in science and natural history museums, this fundamental common denominator of all shows can be more difficult to recognize. Even the long-standing and commonly held distinction within museums between "natural specimens" and "artifacts" on closer inspection becomes rather difficult to maintain, because biological specimens are thoroughly modified from their "natural" state (one might argue, more so than any other kind of exhibit), and so are in fact "artifacts" too.[4] No matter what its subject then (including science), every museum exhibit is, as Hilde Hein has put it, "a work of art of sorts, newly minted for aesthetic contemplation."[5] Stated more prosaically, it is crucial to recognize that there is a non-trivial aesthetic component to creating any exhibition, one which starts with the seemingly superficial "look" of an exhibit, but which almost immediately concerns the ideas behind it.

The most easily recognized exhibit type in which what is visually perceived leads on to its fundamental meaning and significance is, of course, the image or picture. And this leads to a third introductory point that I want to stress, namely the particularly and peculiarly effective role that images can play in exhibitions. A commonly enough understood aspect of images is that they share their workload between an explanatory and inspirational impact, containing both parcels of knowledge and elements of wonder; or, in other words, that they can be considered part science and part art. For this reason, images provide an ideal neutral ground (one that is not monopolized by either domain) for projects that straddle the worlds of science and art. As a consequence, there is enormous, often untapped, potential for exhibitions to exploit images and image making for both their knowledge value and their inherent power to arrest attention. In the first instance, this simply means paying more attention to the aesthetics of science and to the cognitive value of art.[6]

Much of this discussion of art and science meeting on neutral territory lends itself to a vision, eloquently revived in a number of recent works, of the two domains having been falsely and unnecessarily separated. In the giddy company of Leonardo, we are encouraged to remember a time when the two domains were one, and are further invited to reinvent that era, or alternatively find a new third way. My fourth prejudice then is a cautionary one, for my earlier three prejudices have not led me to view the separation of the arts from the sciences as anything other than the inevitable consequence of two hundred years of social and cultural history. Within the Wellcome Trust's extensive recent work to encourage and exploit the links between art and science, I have been struck by few if any signs of a meaningful erosion of the fundamental basis of this separation. Further, in the interminable discussions about the separation of the two domains—a discourse which cannily substitutes words for the works of both art and science—it is hard to find much by way of a truly convincing argument for a single unified culture to replace the existing reality of an intellectual multiculturalism in which science and art are just two strands. But in itself, the division of the "two cultures" does not provide us with a valid reason for concern or despondency. What genuinely is worrying, and indeed undeniably destructive, is the way in which many of the divisions between the two domains have become opaque and impregnable. It is not the existence of the differences between art and science, then, but rather how they are perceived and the fact that it is not often easy to move and exchange ideas between them that has resulted in decided net losses for both art and science. Divisions, then, are inevitable and necessary, but they need not be damaging so long as they remain penetrable and translucent.

Indeed, it is the creative tensions produced by these real and unavoidable differences that then allow sparks to fly when clear lines of communication and navigable paths of cross-fertilization are produced, usually it seems by collaborators from both domains. The sort of interdisciplinary projects alluded to here will be further described later in this paper, but it is worth reiterating now that the core methodology of many of these projects rests on issues of communication between pro-

fessionals who are not used to each other's languages and idioms. And it is in the nature of the most creative and successful of these partnership schemes that the effort, ingenuity and subtlety of mind involved in artists understanding scientists, and vice-versa, is in the end seen less as problems to overcome than as a core strength for the projects actually to draw upon.

All of the foregoing quite possibly makes it redundant to state the last of my introductory prejudices, which is that, in order to present science meaningfully in museums and exhibitions, it has to be treated and presented as part of general culture, not separate from it. Even "contemporary science" has to be seen and understood as existing within an historical and social context. As a number of the contributors to the "Here and now" conference held at the Science Museum in London in 1997 (the immediate forerunner to this volume and conference) make abundantly clear, there is no way of keeping history out of even the most up-to-date science exhibitions. But whereas intelligently self-reflective shows work with this fact, others are simply trapped by it, and as a consequence end up looking dated even when they first open. Thus, while they do much to make science seem like an activity rather than a collection of past ideas, science centers in particular often give no sense of the history upon which the phenomena they examine is based. Ironically, though, many of the scientific ideas that they explore were first discovered a century or more ago. Nonetheless, visitors are commonly given no sense at all that there is a mountain of human endeavor that has preceded and prepared the way for the phenomena and the particular display that they are looking at. As any number of commentators have pointed out, this omission significantly weakens the nature of what are otherwise often very engaging explorations of phenomena and ideas.

In sum, then, this list of presuppositions amounts to an argument for an experimental (essay-like) approach to exhibitions that takes seriously the "artistry" (at the level both of visual stimulus and intellectual content) involved in their production, that fully exploits images as a meeting ground for art and science, and that still recognizes the natural boundaries between the two realms while seeking creatively to permeate them, all pursued in the context of tackling science as part of, rather than separate from, culture. Indeed, it seems to me that art and aesthetics are almost inevitably part of any science/history of science show, and that considerable strength can be gained from embracing and exploiting this fact. The second half of this paper will endeavor to flesh out these contentions with more instances of the artistry of presenting science and its history. Four strategies by which this fertile entanglement between art and science can be mined will be examined: first, the use of art as a user-friendly "vehicle" for science and its history; second, the introduction of artworks as "foils" in scientific exhibitions; third, the idea of art "framing" science; and, fourth, the possibilities inherent in artists reinventing the experimental methodology.

ART AS A "USER-FRIENDLY" VEHICLE

The first method by which art can be employed in science museums boils down to a fairly pragmatic concern with using it as a way of interesting people in science through a medium that much of the public still finds more accessible than undiluted science. There is, however, a clear danger that this approach brings with it—namely that the art ends up doing little more than paying decorative witness to the power of science, with the result that the artwork appears tame and derivative and the science remains obscure. But, when approached imaginatively by gifted scientists, artists and other image makers, this approach can nevertheless be extremely rewarding, and equally important, can reach quite new audiences.

One of the ventures that has recently been undertaken at the Wellcome Trust, inspired initially by the public response to an exhibition program that sought freely to mix contemporary art and science, has been the setting up of a funding initiative for collaborative projects between artists and scientists. In the first two years of this scheme (run under the title sci~art), the Trust has funded a dozen such projects, with widely varying outcomes. Selected on the basis of written applications followed by interviews, and judged by a panel of

experts (in science, art and other domains), projects have been awarded grants in the region of £10,000 and £25,000 each.

One of the six winners awarded money in the first year of the scheme was a project entitled "Primitive Streak," which provides an extremely fine example of this "art as user-friendly vehicle" methodology. Though much else has resulted from this collaboration between a scientist and artist, the primary goal was initially to reconsider and to represent some fundamental scientific ideas through an extremely unusual aesthetic medium. The project involved the sisters Kate and Helen Storey (an Oxford University embryologist and a well-known fashion designer, respectively), who worked together on a fashion show *cum* exhibition that explored and modeled in the form of dynamic and arresting dresses the embryological understanding of the first thousand hours of human existence.

Along with the core knowledge and understanding of the project partners and their inherent skills of communication, the raw data for the project were scientific images produced by a variety of visual technologies, which, as it were, visually told the story of the first month and a half of human life. (Figure 1) At the end of the collaborative process there emerged a series of breath-taking forms that only loosely fit the conventional notion conjured up by the terms *clothes* and *frocks*. (Figure 2 and Figure 3) Thus a "spinal column dress" was made out of red silk printed with patterning derived from a DNA sequence run from a laboratory gel, an ensemble that recalled the molecular basis of embryonic development. It also featured a female spine cast in resin and hand-plated with aluminum foil, which incorporated some eight thousand fibre optic endings, representing nerve processes that extend throughout the body. Another exhibit was a "Heart tube hat," used to show two blood vessels coming together with the development of a diaphragm beneath them, characteristic of the early formation of a human heart. These clothes—products of the collaboration—have subsequently been presented through a very popular international exhibition. (Figure 4)

In this project, the content was by and large embryological and the medium of presentation *haut couture*. But by virtue of the somewhat bizarre nature of the link (or rather gap) between these worlds, and the sheer beauty of the products (the dresses), this project was able to build on the strengths of both worlds, rather than falling between their respective stools. And in terms of audiences, the scientific ideas dealt with in these art forms, or at least the intrigue and curiosity that they have inevitably provoked, has reached an audience sector (adolescent women, for example) that it is very difficult to imagine would have, even with

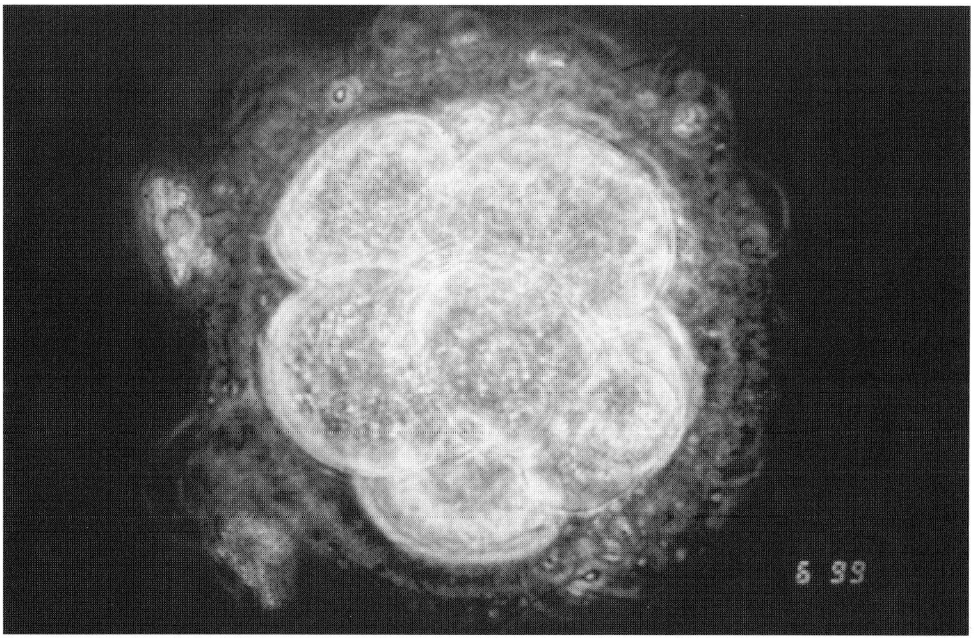

FIGURE 1. Cells dividing in an early human embryo. *Primitive streak.*

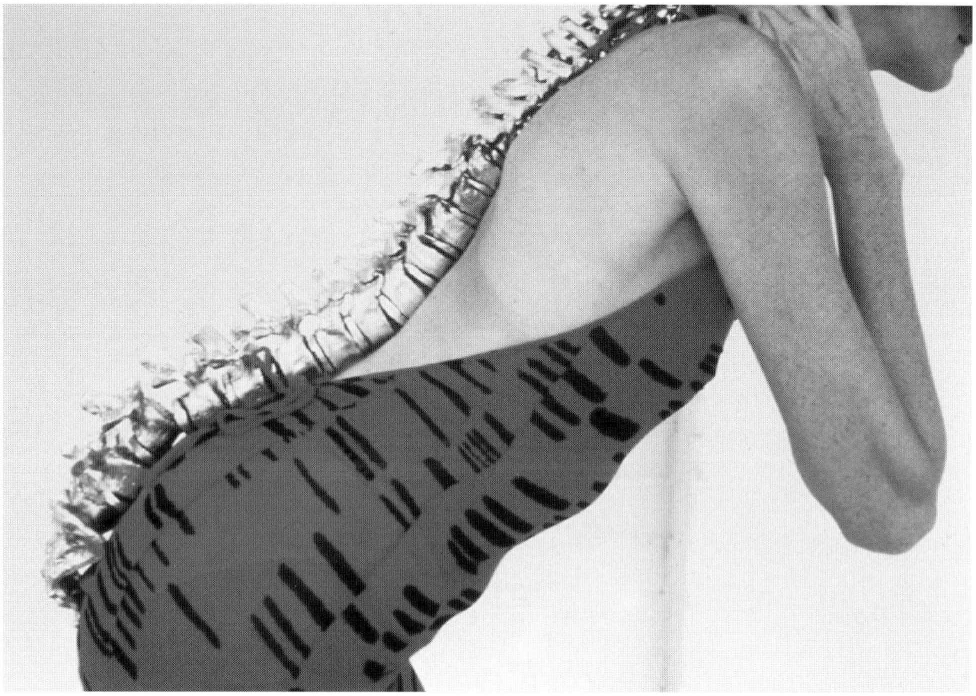

FIGURE 2. Spinal column dress. ***Primitive streak.***
Photo: Jason Lowe

FIGURE 3. Heart tube hat. ***Primitive streak.***
Photo: Jason Lowe

copious amounts of advertising guile, been encouraged to visit a straightforwardly didactic exhibition devoted to the same scientific subject.

ART AS "FOIL"

Another strategy for employing art in science exhibitions touches on the very heart of the exhibition maker's art. For the skill of placing objects, images and texts next to each other in order to make comparisons, suggest contrasts and generally raise issues—in order, that is, to make things talk to each other in the languages of scale, age, value, beauty, significance, not to mention various other emotional and political vectors—is undoubtedly her or his most powerful tool. However, because of the classificatory roots of much museum work, and the consequent habits of abiding by rigorous sensibilities about "what does and does not go with what," this skill of hanging an exhibition is far less masterfully employed than it should be, and certainly so within science exhibitions.

But as, for example, a series of projects undertaken by the American artist and curator Fred Wilson has made clear, the choice and placement of

FIGURE 4. ***Primitive streak*** **exhibition as shown at the Institute of Contemporary Art in London.**

Photo: Helen Storey

exhibits can elevate the often purely pragmatic issues of what to put where to a level of high intellectual audacity. Thus, for example, in his "Mining the Museum" exhibition in Baltimore in the United States, Wilson, under the general heading of "Metalwork 1793–1880," filled a showcase with ornate polished silver goblets, tankards, jugs and the like. Prominent in the center of this display was a rusty hoop, bolt and chain, which, for those who did not immediately recognize it, was identified by a label as a slave's shackle. In one bold stroke, then, the comfortable and polite economy of fine craftsmanship made for elite appreciation conjured up by the silverware was interrupted by the politics of oppression and the sentiments of human misery embodied in a rudely formed piece of iron. Juxtaposition in exhibitions has the power to do this, and artworks and the artistic imagination can, it is clear, be particularly effective at making the choice and placement of objects "say provocative things."[7]

In a very different fashion, an exhibition held at the Wellcome Trust in London in 1996 also sought to place the theme of juxtaposition at its core. *Materia medica: a new cabinet of medicine and art* was based on the collaboration between the curators (Martin Kemp and Ken Arnold) and a dozen British artists, each invited to present a mixture of their own work alongside material relating to the history of medicine, drawn from the collections held in the Wellcome Institute Library and objects now kept on "permanent loan" at the Science Museum.[8] (Figure 5)

The intellectual inspiration for the project came in part from the very early history of museums, that is from the Renaissance cabinets of curiosity that stand at the very beginnings of the modern museum movement. (Figure 6) For these collections of (from our point of view) dizzying and at times comically heterogeneous material—from art, science, and nature, but also religion, folklore, proto-ethnography, magic, and indeed plain misunderstandings—provided a cue to the potential intrigue and power of bringing together exhibits that simultaneously possessed a thread of continuity and a jolt of discontinuity. The language of physical juxtaposition for the sake of visual eloquence invented in these Renaissance cabinets has been at the heart of displays, exhibitions and museums ever since.

One example of works shown in the *Materia medica* exhibition brings out some of the potential embedded in its core methodology. In the late 1980s, the Scottish artist John Bellany was seriously ill. His liver was collapsing, and at times it seemed to be taking the rest of his body and mind with it. Initially treated in St. Thomas' Hospital, his deteriorating condition made a liver transplant the best medical option. After exhaustive tests, the

sentations, Meaning, Organization, and Practice, ed. Steven Laurence Kaplan and Cynthia J. Koepp (Ithaca, NY, London: Cornell University Press, 1986), pp. 229–257. John R. Pannabecker, "Representing Mechanical Arts in Diderot's Encyclopédie," *Technology and Culture*, 39 (1998): 33–73.

18. Michael Polanyi, *Personal Knowledge* (Chicago: The University of Chicago Press, 1958). Michael Polanyi, "Tacit Knowing," in *The Tacit Dimension*, Michael Polanyi (Gloucester, MA: Peter Smith, 1983), pp. 3–25.
19. See, for example, Mario Biagioli, "Tacit Knowledge, Courtliness and the Scientist's Body," in *Choreographing History*, ed. Susan Leigh Foster (Bloomington: Indiana University Press, 1995), pp. 69–81.
20. For criticism of a static conception of knowledge, see Ludwik Fleck, "Das Problem einer Theorie des Erkennens," in *Ludwik Fleck. Erfahrung und Tatsache. Gesammelte Aufsätze*, ed. Lothar Schäfer and Thomas Schnelle (Frankfurt a.M.: Suhrkamp, 1983), pp. 84–127.
21. See H. Otto Sibum, "Reworking the Mechanical Value of Heat: Instruments of Precision and Gestures of Accuracy in Early Victorian England," *Studies in History and Philosophy of Science*, 26, no. 1 (1995): 73–106. As well as, H. Otto Sibum, "Les Gestes de la Mesure. Joule, les Pratiques de la Brasserie et la Science," *Annales—Histoire, Science Sociales*, 4–5 (1998): 745–774.

6. The historian of physics Jürgen Teichmann has initiated several projects. See, for example, Jürgen Teichmann, "Die Rekonstruktion historischer Modelle und Experimente für den Unterricht—drei Beispiele," *Physik und Didaktik*, 4 (1979): 267–282. Jürgen Teichmann, "Das historische Experiment im Physikunterricht—wissenschaftstheoretische Betrachtungen und praktische Bedeutung," in *Wissenschaftsgeschichte und naturwissenschaftlicher Unterricht*, ed. Michael Ewers (Bad Salzdetfurth, 1978), pp. 185–212. See also note 15.
7. Falk Rieß, "Erkenntnis durch Wiederholung—eine Methode zur Geschichtsschreibung des Experiments," in *Experimental Essays—Versuche zum Experiment*, eds. Michael Heidelberger and Friedrich Steinle (Baden-Baden: Nomos Verlagsgesellschaft, 1998), pp. 157–172.
8. For the reconstructions described, see Uwe Mehrle, Jörg Meya, Falk Rieß, Gerhard Rohlfs, and Heinz Otto Sibum, eds., *From the Electrical Machine to the Theory of Ether Apparatus and Models in History of Electricity from the 18th to the 19th Century* (Oldenburg: University of Oldenburg, 1987). Among recent qualifying works in the Department of Physics, see the diploma work of Christian Sichau, *Der Joule-Thomson-Effekt: Der Versuch einer Replikation* (University of Oldenburg, 1995). See as well the diploma work of Jan Frercks, *Die Zahnradmethode zur Messung der Lichtgeschwindigkeit: Replikation von Fizeaus Experiment* (Oldenburg: University of Oldenburg, 1996).
9. See, for example, the different approaches by Stillman Drake, "Renaissance Music and Experimental Science," *Journal of the History of Ideas*, 31 (1970): 483–500; Stillman Drake, "Galileo's Experimental Confirmation of Horizontal Inertia: Unpublished Manuscripts," *ISIS*, 64 (1973): 291–305; Thomas B. Settle, "An Experiment in the History of Science," *Science*, 133, no. 1 (1961): 19–23; Thomas B. Settle, "La rete degli esperimenti Galileiani," in *Galileo e la scienza sperimentale*, eds. P. Bozzi, C. Maccagni, C. Olivieri, and T. B. Settle (Padova: Dipartimento di fisica Galileo Galilei, 1995), pp. 11–62; Svante Lindquist, *Technology on Trial: The Introduction of Steam Power Technology into Sweden, 1715–1736* (Uppsala: Almqvist & Wiksell International, 1984); Lawrence Principe, " 'Chemical Translation' and the Role of Impurities in Alchemy: Examples from Basil Valentine's Triumph-Wagen," in *Ambix*, 34 (1987): 21–30.
10. Steven Shapin and Simon Schaffer, *Leviathan and the Airpump: Hobbes, Boyle, and the Experimental Life* (Princeton: Princeton University Press, 1985), p. 16.
11. Peter Galison, *How Experiments End* (Chicago: University of Chicago Press, 1987). Compare also Simon Schaffer, "Where Experiments End: Tabletop Trials in Victorian Astronomy," in *Scientific Practice: Theories and Stories of Doing Physics*, ed. Jed. Z. Buchwald (Chicago: University of Chicago Press, 1995), pp. 257–299.
12. David Gooding, *Experiment and the Making of Meaning: Human Agency in Scientific Observation and Experiment* (Dordrecht: Kluwer, 1990).
13. On laboratory notebooks, see H. Otto Sibum, "Narrating by Numbers. Keeping an Account of Early 19th Century Laboratory Experiences," forthcoming in *Reworking the Bench*, eds. F. L. Holmes, J. Renn, and H.-J. Rheinberger.
14. The examples discussed are mainly based on my investigations of James Prescott Joule's experiment to determine the mechanical equivalent of heat. James Prescott Joule, "On the Mechanical Equivalent of Heat," in *The Scientific Papers of James Prescott Joule*, ed. The Physical Society of London (London: Taylor and Francis, 1884), pp. 298–328. For detailed accounts about the experimental performance, see H. Otto Sibum, "Reworking the Mechanical Value of Heat: Instruments of Precision and Gestures of Accuracy in Early Victorian England," *Studies in History and Philosophy of Science*, 26, no. 1 (1995): 73–106.
15. On some results of this project, see Thomas Eisenblätter and Jürgen Teichmann, *Was nützen historische Modelle und Rekonstruktionen*, Second Edition (München: Deutsches Museum, 1997).
16. We see here possibilities for co-operation with museums and academic departments. The Departments of History of Science at the University of Cambridge and Harvard University, as well as the History of Science Museum at Oxford University already work very intensively toward the integration of their instrumental resources with the practical education of the historian of science. See also two exhibitions which represent the results of such an interdisciplinary collaboration: *Empires of Physics. A Guide to the Exhibition*, eds. J. Bennett, R. Brain, K. Bycroft, S. Schaffer, H. O. Sibum, and R. Staley (Cambridge: Whipple Museum of the History of Science, 1993). *The Physics of Empire*, ed. R. Staley (Cambridge: Whipple Museum for the History of Science, 1994). As well as, *1900. The New Age*, eds. J. Bennett, R. Brain, K. Bycroft, S. Schaffer, H. O. Sibum, and R. Staley (Cambridge: Whipple Museum for the History of Science, 1994).
17. Cynthia J. Koepp, "The Alphabetic Order: Work in Diderot's Encyclopédie," in Work in France. *Repre-*

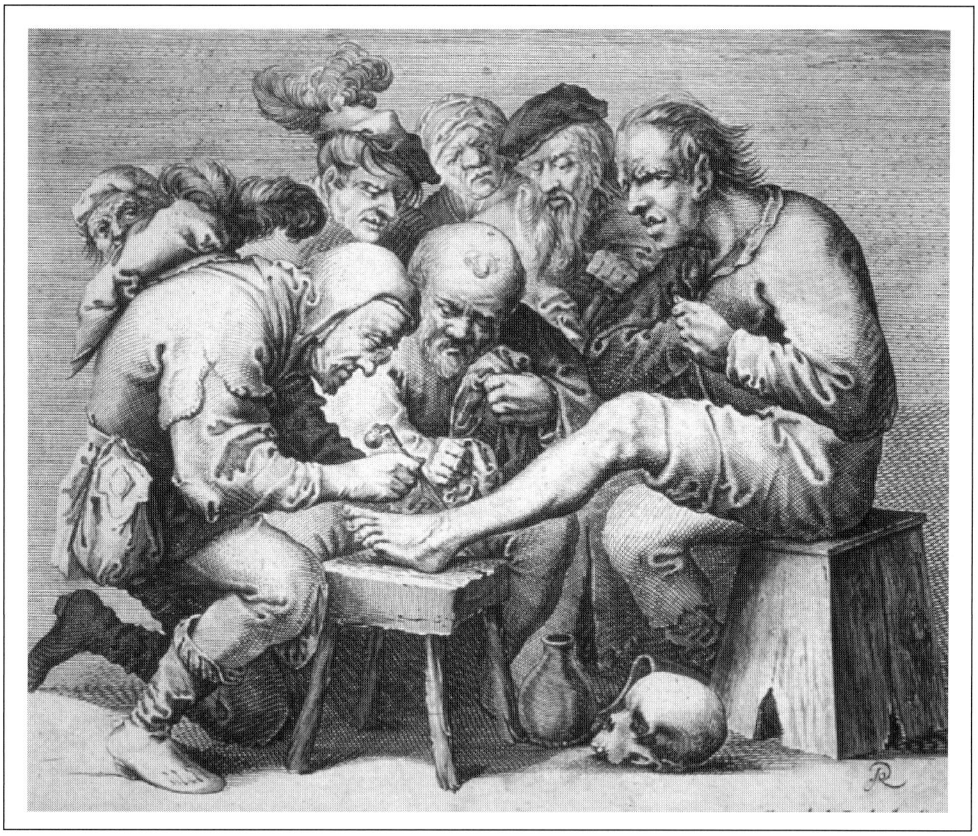

FIGURE 7. A rural surgeon treating an elderly man's foot. Etching by Adolphe Théodore Jules Martia Potemont after David Teniers, the younger. Mid-nineteenth century.

Courtesy Welcome Library

plant operation. (Figure 8) The audacious medical event that Bellany survived gave him the sense that the door on his life had been left slightly ajar, and that his task thereafter was to open it and peer out into his new world. Gradually, the concerns of his work turned from the internal focus reflected in this work to subjects outside himself.

Old versus new, public as opposed to private, art from a patient next to art of an operation: these were some of the axes along which this coupling of exhibits worked. This single example of an exhibit pairing is indicative of how the exhibition *Materia medica* attempted to exploit the museum space as one ideally suited to relating on the one hand the "art of medicine" to medical significant art, and on the other the old and the new.

FIGURE 8. "Self-portrait: diptych." Etching by John Bellany. 1988.

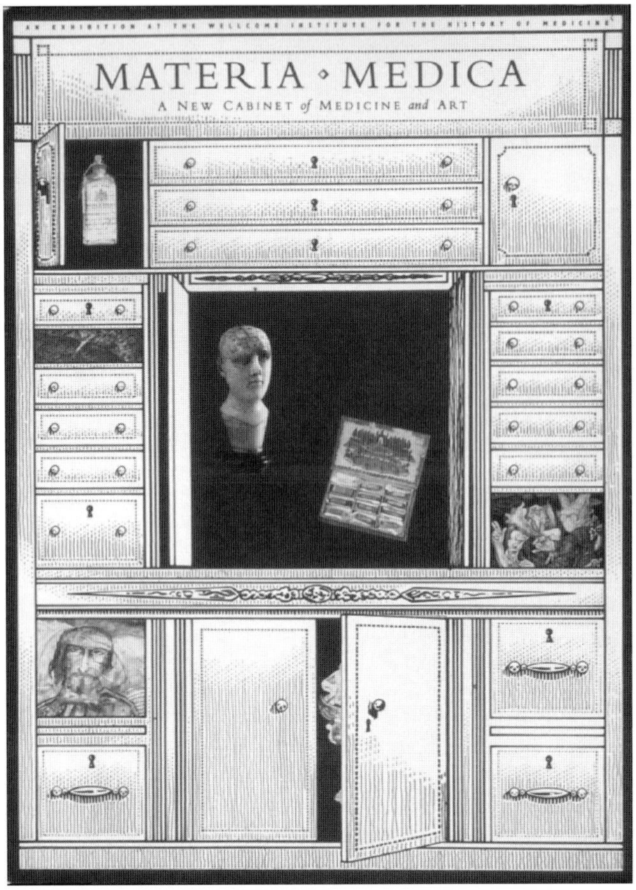

FIGURE 5. ***Materia medica: a new cabinet of medicine and art.***

operation was finally carried out in the spring of 1988 in Addenbrooke's Hospital. In every conceivable sense, the successful operation gave Bellany a new lease on life. Almost as soon as he regained consciousness, he started drawing again; his art both helped him to deal with the pain of recovery and provided an essential tool in the task of reconstructing his life. This extraordinary encounter with modern medicine both allowed Bellany to continue practicing his art and also strongly influenced his approach to it. It was this extraordinary story and the broader themes of the parallel experiences of artist and patient that it embodied that were reflected in the mixture of Bellany's own work and his choice of historical material shown in the exhibition.

Thus a line engraving by Pieter Jansz Quast (Amsterdam, 1606–1647), showing an audience watching a surgeon operate on a patient's foot, was to Bellany fascinating for its clear depiction of the way in which early surgical operations could almost become theatrical performances. (Figure 7) Shown alongside it was Bellany's own etching "Self-portrait: diptych" (1988), a pair of images which seem to capture Bellany's mood and self-perception during his recovery from the liver trans-

FIGURE 6. Interior of Ole Worm's museum in Copenhagen. Frontispiece to his *Musei Wormiani Historia. . .* (1655).
Courtesy of Wellcome Library

EXHIBITION AESTHETICS

The third methodology for using art that I want to explore has already been alluded to, namely exhibition aesthetics. Using the look of an exhibition (everything from the gallery setting and ambient light levels to styles of frames and methods of labeling) can often provide a rich and nuanced way (borrowing Stephen Greenblatt's terminology) to both parade the "wonder" and channel the "resonance" of individual exhibits—that is, of using a given exhibit to stop visitors in their tracks and then provide them with further understanding through a series of associated objects, images and ideas.[9]

The idea that the manner in which objects are shown can profoundly influence how they are viewed and understood is, by now, a fairly well accepted museological notion. A quick glance at the history of display styles from earlier eras of museum presentations suggest just how much meaning can be derived from techniques of exhibition aesthetics. In the world of art, for example, the early 19th-century aesthetic for crowding pictures together and thereby "creating an ideal organic unity" was later replaced by a style of display which provided larger spaces between works and a more didactically determined sequence in order to focus attention on individual works of art. Thus the earlier organic mass was divided up into sections and spread out in a more linear fashion, with the consequent emergence of a "story-line." More recently this shift in aesthetics has been taken a stage further, with artworks now most frequently presented in enormous empty white spaces in a seeming effort, as Carol Duncan has put it, to make the museum seem sacred and thereby encourage the viewers to "commun[e] with immortal spirits." In natural history museums, however, the original fashion for taxonomic arrays of specimens grouped by type were, in the nineteenth century, first augmented by specimens preserved and mounted in more life-like poses and then by plants and animals inserted in dioramas and panoramas that evoked "natural habitats," and through which, eventually, visitors were invited to walk, as if momentarily transported to those unfamiliar "natural" locations. Then again, in the case of ethnographic objects, which for much of their early history shared the fate of natural history collections, more recent exhibition aesthetics have removed them from this abstract classificatory context and instead inserted them first into ideal recreations of timeless and inconceivably "other" places, and then more recently still re-evaluated them as "art" objects worthy of the white cube and halogen spot-light "high art" treatment. Recent scholarly and monographic exhibitions, particularly those pioneered in North America, have produced an exhibition aesthetic whose aim is to embody a carefully scripted narrative, so that a sequence of objects shown in context is expected to encourage visitors to follow the curator's own thinking from one complex of ideas to another. Working in a rather different direction, another group of recent exhibitions have self-consciously paraded the multiple meanings of object collections. The "Art/Artifact" exhibition curated by Susan Vogel at the African Center in New York in 1988, in which African artifacts were presented in two very different fashions as alternately "art" and "anthropology" (expressed as very different exhibition aesthetics) is axiomatic of this new type of concentration on the way exhibits are presented.[10]

Two hundred or more years of the history of museum aesthetics then backs up the fundamental sense that how things look greatly influences what stories are attached to them. This point about the significance of exhibition aesthetics was brought home to me in a particularly compelling way during a recent visit to the pioneering Exploratorium in San Francisco. The Exploratorium, as its subtitle proclaims, it not only a museum of science, but also of "art and perception," and it is the fashion in which a number of the very arresting contemporary artworks on display there are presented that intrigued me. Since its founding, the museum has involved artists at a number of levels of its operation, including an active artist-in-residence program, through which a good number of the exhibits permanently on display have been produced. The museum's website lists about fifty.[11] (Figure 9 and Figure 10)

What is so striking about the presentation of these artworks, especially for someone used to seeing them in art museums, is just how easy it is not to realize that they are "art" at all. Careful inspection of the captions usually does reveal a named individual credited as "artist," but often one feels that

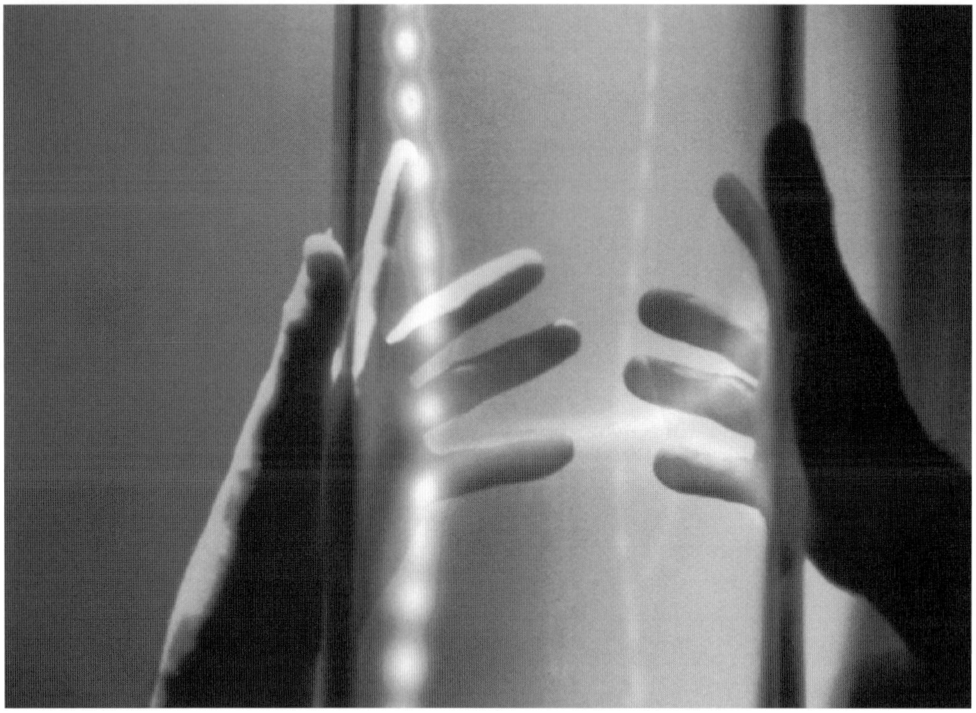

FIGURE 9. "AM lightning" by Bill Parker, as shown at the Exploratorium.
Photo: Nancy Roger, 1974

FIGURE 10. "Sun painting" by Bob Millar, as shown at the Exploratorium.
Photo: Nancy Roger, 1983

this is as much if not more for the benefit of the artist than the visitors. There are decidedly no frames, no blank white-washed walls, no hushed halls, no uniformed guards: in short, no art museum to tell you that there *is* art there. Were most visitors to the Exploratorium asked what their favorite artwork in the museum was, my suspicion is that they would simply deny having seen any.[12]

As Hilde Hein, the "biographer" of the Exploratorium, has explained, the difference between "a work of art and an instrument of science is that the former draws attention to itself as the locus of an aesthetic experience, but the latter, even if it is beautiful, directs attention away from itself to a process or phenomenon."[13] What is interesting about the display of artworks at the Exploratorium is that they too are presented as "instruments."[14] And what certainly shines through this methodology is the extraordinary skill with which artistic imaginations can guide our understanding of nature and natural phenomenon.

Moreover, this particular exhibitionary aesthetic is thoroughly convincing in its presentation of both science and art as processes of discovery and ways of dealing with the natural world. And this is important, for it provides a powerful reminder of the

need to be on one's guard against the myth, still perpetuated by some more traditional science museums, that science is nothing more than "a set of accomplishments that have already been achieved."[15] This significant point admitted, it nonetheless remains crucial to grasp that both science and art are as much products as they are processes—both have for most of their modern histories, at least, been full to bursting with objects of all sorts, and with the myriad meanings that accompany them.[16] The problem with the presentation style of the Exploratorium, then, is that it provides the visitor with little if any insight into the nature of the *material culture* of either science or art. And as a fair number of recent histories have stressed, one cannot understand either art or science without thinking long and hard about this aspect of both cultures. All the Exploratorium's objects (and one should be in no doubt, there are literally hundreds and thousands of objects to be seen there) are treated as if they are more or less transparent. One is continually and consistently encouraged to look *through* them onto the phenomenon they are meant to illuminate. The pipes, mirrors, computers, lights, bottles, switches, blocks, models, telephones, and so on and so forth are all made to disappear.[17]

There is not space in a short essay such as this to expand the point about just how much insight into the history and practice of science is provided by an analysis of its material culture. Luckily, many of the other contributions to this volume make the point far more tellingly than I could. I will therefore restrict myself to the description of a single exhibition presented at the Wellcome Trust in London that has, as it were, deliberately worked in the opposite aesthetic direction to that of the Exploratorium. For *Inside information: imaging the human body* was aimed explicitly at bringing out the object-like aesthetic qualities of elements of modern science that are, while they are still in the scientific domain at least, commonly simply seen-through.

The exhibition, taken from the book of the same name by William Ewing, took a selection of recent medical images and, in short, presented them as if they were artworks. Here then was an attempt to take scientific images which are often used instrumentally—"read" in order to decipher the fundamental biological reality that they point toward—and turn them into immutable objects that visitors were invited to treat in aesthetic terms before moving on to what they "meant" scientifically. By the fashion in which the pictures were presented—using such "art" conventions as naming the image-maker and giving the pictures "titles"—the show attempted, as it were, to rescue some images that, in their professional context, are commonly looked through and turn them instead into objects of beauty, suggestive of aesthetic sensibilities at work in an area of "visual science." (Figure 11) Subsequent exhibitions of visually arresting biomedical images presented in the same gallery have exploited the same methodology. (Figure 12)

What such exhibition "tricks" reinforce is the point made frequently enough about the nature of scientific images, indeed images in general: that rather than just visual records of what is there, they should instead be understood as "diagrams" with encoded information, which, either explicitly or implicitly, are accompanied by visual conventions adopted by both creator and user. This point is crucial, because the scientific content of an image can only successfully be transmitted if creators and viewers share the same interpretation of what they are looking at. In fact, scientific images inevitably, of course, carry with them a series of contexts, from the technical apparatus of captions, scales and codes that guide the viewers "reading" and understanding of them, to the broader cultural and social contexts that constitute a framework of belief. Exhibition aesthetics, then, have the power both to make art seem like an instance of scientific process, and to turn scientific tools into objects of aesthetic contemplation. It might be argued that both these extremes are evidence of the power that the exhibition maker's art has to deceive its audience, but the general point unarguably stands that the way in which exhibits are shown—be they science or art—can have almost as great an impact on how an audience understands them as the "raw" object itself.

ART AS "IMPOSSIBLE VISUAL EXPERIMENTS"

The fourth and final example of a methodology by which art can be usefully employed by science exhibition makers relates to an activity that has come

FIGURE 11. ***Inside information*** **exhibition as shown at the Two10 Gallery, Wellcome Trust, London, 1996.**

to be assumed to be the exclusive domain of the sciences: experimentation. But numerous contemporary artists have been keen to reinvent the "experiment" in highly suggestive ways. Thus the British artist Vered Lahav describes his own methodology as that of a "visual scientist" doing "impossible experiments." And it is under the same banner of experimentation that the more forward-thinking galleries involved in this area of exhibitions, echoing some ventures undertaken in the orgy of experimentation that surfaced during the 1960s, have self-consciously performed as laboratories, "testing grounds for reality" as one commentator has described them.[18]

One example, a beneficiary of the 1998 Wellcome Trust's science and art funding initiative (sci~art), of a project that embodies just such a "visual experiment" might suffice to make clear just how rich is the potential of working in this fashion. Like many of the sci~art winners, "The Artist's eye" project has produced significant outcomes in the various forms of new artwork, interesting sci-

FIGURE 12. ***Biomedical image awards*** **exhibition. Two10 Gallery, Wellcome Trust, London, 1998.**

entific results and public demonstrations, in this case an exhibition. It also embodies two other elements of the experimental methodology employed by the most successful of such projects which seem crucial to providing the necessary energy for the success they have reaped. First, "Artist's eye" was driven by collaborators who were able to embrace the risks inherent in finding and then, as fully as possible, understanding each other's interests, skills, perspectives and ways of working. Second, the project incorporated a range of abilities and enthusiasms shared between the partners for initially encouraging wild improbable dreams and then, just as importantly, insisting that they come back to earth as real and concrete—that is, imaginative and innovative "what ifs" followed by feasible and practical "therefores."

Co-ordinated by film-maker John Tchalenko, the "The Painter's eye movements" project drew on expertise in visual perception shared by a practicing artist (Humphrey Ocean) and a group of leading cognitive and brain scientists (including Chris Miall). It also used state-of-the-art biomedical equipment—namely an fMRI scanner and an eye tracker. The focus of the study was the processes by which the external world is transformed through an artist's vision into a picture (a portrait in this case). The methodology simultaneously evolved and employed by the project team was a minute and elaborately layered set of observations of the artist at work.[19]

Part of the project involved drawing exercises (Figure 13) done while the "patient's" brain was being scanned using the fMRI (functional magnetic

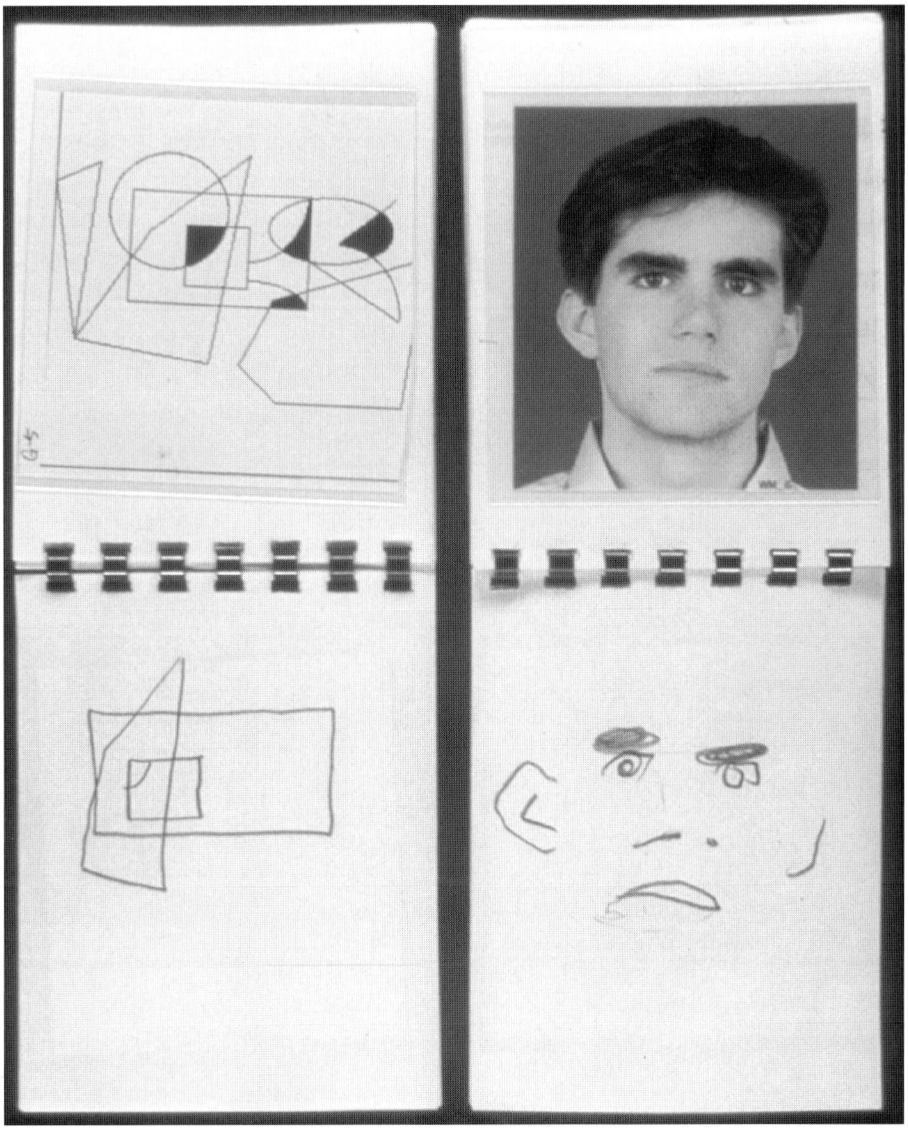

FIGURE 13. Drawing exercise from "The Painter's eye movements" project. The drawings at the bottom were drawn by Humphrey Ocean from photographs in 30 seconds while undergoing fMRI scan.

resonance imaging) technique. This equipment allows the detection of increased blood flow in particular regions of the brain, and thereby the identification of areas of the brain specifically activated by doing particular tasks. The project compared the results of such a task being undertaken by the artist Humphrey Ocean and some "non-experts." The results of this highly innovative, though not exhaustive, study suggest that experts might actually use different parts of their brains than those drawn on by non-experts performing the same tasks.

Another aspect of the project involved the very close study of eye and hand movements by Humphrey Ocean while he was producing a pencil portrait. In this case, the artist at work was asked to wear an eye-tracker, while a digital camera simultaneously recorded the drawing as it emerged on the page (Figure 14). This mix of equipment allowed scientists to record very accurately the artist's eye movements during the process of drawing and relate these to his employment of his pencil to render the drawing. The results were rich and fascinating, suggesting any number of interesting theories about how the artist's eye captures an impression of what he is studying and how this is then translated into a series of graphic records made on the paper.

Here then was a multifaceted project that shed much light on such questions as how an artist's observations of his model are processed in the brain, and how he reproduces with his hand what his eye has seen. At a more general level, the project also added fresh insight into such issues as what distinguishes an expert's and an amateur's approach to producing pictures, and, further, the processes involved when any of us look at portraits, or indeed any picture. Not just science applied to artistic situations, the project employed a considerably more holistic outlook than one could anticipate emerging from any narrow single-discipline study. What makes it relevant to the issues under discussion here is that the project also generated the production of works of considerable artistic merit—a portrait that Ocean himself considers one of his finest, and a highly original documentary film. (Figure 15) Further still, all of this (the science, the art and the film-making) have been condensed into a very unusual and successful exhibition shown initially at the National Portrait Gallery in London—an exhibition packed with as much food for thought as feasting for the eye.

The "Painter's eye" captures perfectly the potential presented to projects that overtly and creatively mix the sensibilities, disciplines, inspiration and insights of art and science that have been the subject of this essay. For, in part, it presents scientific con-

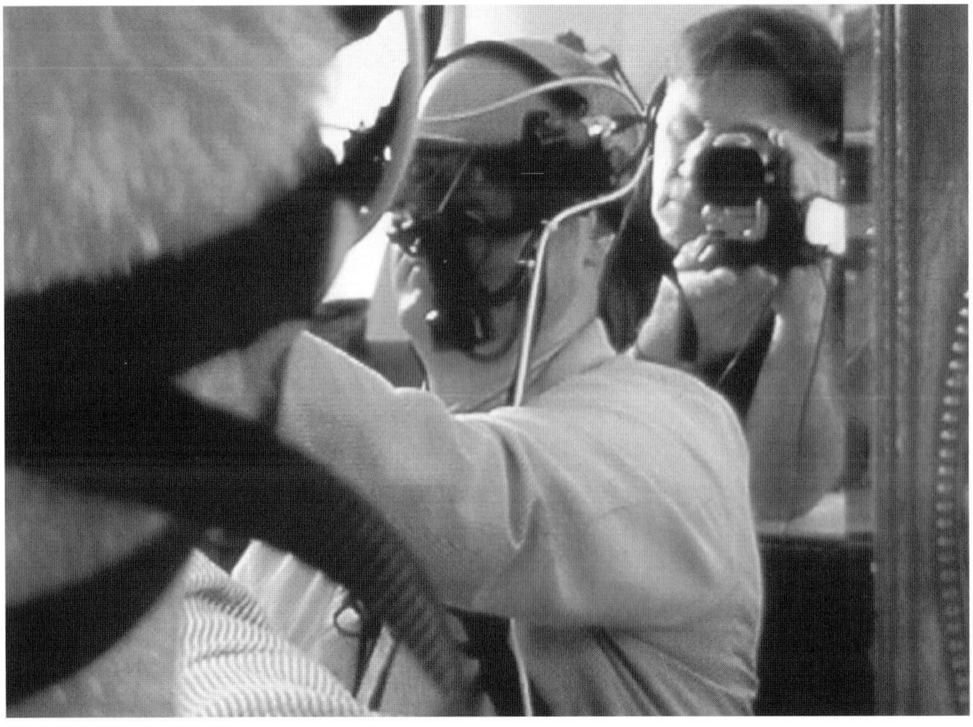

FIGURE 14. Humphrey Ocean at work and under observation.

Still from video by Dr. John Tchalenko.

FIGURE 15. "Nick" by Humphrey Ocean. Pencil on paper. 79.5 x 59.5 cm 15 July 1998. (Detail)

cepts through an artistic medium (and almost undoubtedly to new, uninitiated audiences). Just as significantly, it also introduces art (in this case, that of a human face) as a "foil" for the rigors of the science. Furthermore, it frames much of the *real* new science (not insights that have been made elsewhere and then incorporated into the project) generated by the project within an interesting aesthetic medium—in this case through John Tchalenko's film and the innovative exhibition. And finally, it manages to do all this by mounting an "impossible visual experiment." One is tempted to conclude that this is fancy science and factual art at its best.

NOTES

1. Quote from interview conducted at Montreux, Switzerland in September 1966. (Published in *Wisconsin Studies in Contemporary Literature*, vol. VIII, no. 2, spring 1967).
2. See Ken Arnold, "From Collections of Wonder to Cabinets of Taxonomy: a contemporary view of early English museums," in *Jahrbuch für Geschichte und Theorie der Biologie /, herausgegeben . . . in Verbindung mit der Deutschen Gesellschaft für Geschichte und Theorie der Biologie*, ed. Anke te Heesen. (Based on paper given at "The natural history cabinet and its function in the eighteenth century" conference, Potsdam, February 1998.) [Forthcoming—Berlin: Verlag für Wissenschaft und Bildung.]
3. Greenberg, Ferguson and Nairne have described exhibitions within the art world as *the* "primary site of exchange in [its] political economy," in *Thinking about exhibitions*, eds. Reesa Greenberg, Bruce W. Ferguson, and Sandy Nairne (London, New York: Routledge, 1996), pp. 297–305. For Savitsky Gallery, see Valery Petrovich Sazonov, "The one-picture gallery," in the same volume, pp. 231–250.
4. This point is made by R. Y. Edwards, "Research: a museum cornerstone," in *Museum collections: their roles and future in biological research*, ed. E. H. Miller (N.P.: Ministry of the provincial secretary and government services, 1985), pp. 1–11, p. 9.

5. Hilde Hein, "The art of displaying science: museum exhibitions," in *The elusive synthesis: aesthetics and science*, ed. Alfred I. Tauber (Dordrecht, Boston, London: Kluwer Academic Publishers, 1996), p. 268.
6. Interestingly enough, many contemporary artists and scientists seem to have identified visual images, and maybe more specifically the technologies and sites of image making, as a fruitful place in which unusual inspiration and even interaction can occur. Pictures and the processes that go into producing them seem to provide a means by which artists and scientists can "talk" to each other. The ethnographer Thomas Cummins has similarly identified pictorial images as "a mutually recognised cultural space" in which the Europeans who first "discovered" the New World and the native peoples already there could begin to translate ideas from one culture to another. There, of course, the similarities between these two encounters end. [Thomas Cummins, "From lies to truth: colonial ekphrasis and the act of crosscultural translation," in *Reforming the Renaissance: visual culture and Latin America, 1450–1650*, ed. Claire Farago (New Haven: Yale University Press, 1995), pp. 152–174.]
7. Even the title of Wilson's Baltimore show carried several messages, for mining was meant in the sense of making it "mine" (i.e., his), of digging in the museum's archives and of exploding the museum. See *Mining the museum: an installation by Fred Wilson*, ed. Lisa G. Corrin with contributions by Leslie King-Hammond and Ira Berlin (New York: The Contemporary, Baltimore in cooperation with the New Press, 1994) and Ivan Karp and Fred Wilson, "Constructing the spectacle of culture in museums," in *Thinking about exhibitions*, eds. Reesa Greenberg, Bruce W. Ferguson, and Sandy Nairne (London, New York: Routledge, 1996), pp. 251–267. In a different context, Peter Vergo has also stressed the potential within exhibitions of using non-verbal material (maps, photos, pictures, diagrams, and other objects) as a means to illuminate, explain or problematize an object on display. See Peter Vergo, "The reticent object," in *The new museology*, ed. Peter Vergo (London: Reaktion Books, 1989), pp. 41–59, p. 53. Finally, James Clifford has made an intriguing argument for the role of juxtaposition in early ethnographic museums, which he argues was directly borrowed from avant-garde surrealist art. See James Clifford, "On ethnographic surrealism," in *Comparative studies in society and history*, vol. 23, no. 4 (181): 539–546.
8. Ken Arnold and Martin Kemp, *Materia medica: a new cabinet of medicine and art* (London: Wellcome Trust, 1995).
9. Stephen Greenblatt, "Resonance and wonder," in *Exhibiting cultures: the poetics and politics of museum display* (Washington and London: Smithsonian Institution Press, 1991), pp. 42–56, p. 42.
10. See Germano Celant, "A visual machine: art installation and its modern archetypes," in *Thinking about exhibitions*, eds. Reesa Greenberg, Bruce W. Ferguson, and Sandy Nairne (London, New York: Routledge, 1996), pp. 371–386, p. 374; Carol Duncan, *Civilizing rituals: inside public art museums* (London: Routledge, 1995), p. 17. Karen E. Wonders, "Bird taxidermy and the origin of the habitat diorama," in *Non-verbal communication in science prior to 1900*, ed. Renato G. Mazzolini (Florence: Leo S. Olschki, 1993), pp. 411–447. Hilde Hein, "The art of displaying science: museum exhibitions," in *The elusive synthesis: aesthetics and science*, ed. Alfred I. Tauber (Dordrecht, Boston, London: Kluwer Academic Publishers, 1996): pp. 276–280.
11. The museum's founder, Frank Oppenheimer, deliberately included art "not just to make things pretty, but primarily because artists make different kinds of discoveries about nature than do physicists or biologists." Quoted on the Exploratorium's web-site, where further information about the museum's arts policy and program can also be found (www.exploratorium.com). See, too, Hilde Hein, *The Exploratorium: the museum as laboratory* (Washington and London: Smithsonian Institution Press, 1990). For further discussion of art at the Exploratorium, see Peter Richards, "From London to Nagasaki—the roots of interactive art @ the Exploratorium" in *Art@Science*, eds. Christa Sommerer and Laurent Mignonneau (New York: SpringerWeing, 1998), pp. 215–224 and *A curious alliance: the role of art in a science museum* (San Francisco: The Exploratorium, 1994).
12. There are some clear benefits to be gained from displaying art without announcing it as such—particularly, maybe, for the contemporary art world. More people are exposed to highly innovative artworks at the Exploratorium than at many a medium-sized art gallery. Moreover, they are probably not the sort of people who would necessarily go out of their way to see contemporary art. And furthermore, they thoroughly enjoy what they see, which might not be the case if the same pieces were displayed in an art gallery. Clearly, just as art can be used to gain new audiences for science (as was my contention when describing the "Primitive streak" dresses above), so science can be used to reach new audiences for art. One might, with a touch of irony, suggest that the

Exploratorium does as much for the public understanding of art as for science.

13. Hilde Hein, *The Exploratorium: the museum as laboratory* (Washington and London: Smithsonian Institution Press, 1990), p. 36.
14. Hein has further described exhibits at the Exploratorium as "vehicles of ideas, meant to drive visitors along a conceptual road from the experience of an exhibit to the comprehension of other experiences and principles bound to it by way of a common concept," which "direct attention back to the world rather than being a refuge from it." [Hilde Hein, "The art of displaying science: museum exhibitions," in *The elusive synthesis: aesthetics and science*, ed. Alfred I. Tauber (Dordrecht, Boston, London: Kluwer Academic Publishers, 1996), p. 273.]
15. Far from a warehouse of dead ideas, the Exploratorium is much more a factory, or better a laboratory, of living ones. A particularly important aspect of the Exploratorium, one which brings out this focus on process particularly elegantly, is the fact that the "machine shop," where many of the exhibits are made, is very much itself one of the exhibits—quite possibly the most interesting one on show. Looking at what goes on there embodies the sense that making is a way of learning. But it also brings humans and their cultural habits, other than those of the individual visitors themselves, back into the frame. And for this visitor at least, it was here watching people working on future exhibits that I found to my relief a reminder that science is indeed as much a part of human culture as it is the external natural world. [See Hilde Hein, *The Exploratorium: the museum as laboratory* (Washington and London: Smithsonian Institution Press, 1990), p. 2 & passim.]
16. Curiously, fine art objects are, sometimes, similarly presented as if they are somehow transparent—i.e., that they invite viewers to look through, or at least be transported through them, into a different realm in which the onlooker can come face to face with pure creativity and genius. For many art museums also pay scant attention to their objects as material culture. The other contemporary context in which visitors are also invited to "look past" the immediacy of the "objects" on show is in the heritage center, where props and fakes are used to evoke the material conditions of a former era. In all three cases, it is not that objects are missing, but rather that their interpretation as objects is not forthcoming.
17. In a curious way, they are used just like sacred objects within a religious context in which strictly speaking the material objects are meant to be used as devices to transport the believer onto a higher plane of understanding.
18. Galleries such as Arts Lab and the Centre for Advanced Creative Study were set up in mid-60s London, and John Kasmin set up his Park Place in New York in 1963 under the subtitle of "Gallery of Art Research, Inc." "Any work of art that was really good, [said Kamin], should be tested." So that, for him, the gallery "was more like a laboratory. It was a place for testing works of art." Another model for the gallery that has been experimented with is the industrial research lab. [Reesa Greenberg, "The exhibited redistributed: a case for reassessing space," in *Thinking about exhibitions*, eds. Reesa Greenberg, Bruce W. Ferguson, and Sandy Nairne (London, New York: Routledge, 1996), pp. 349–367, p. 353. See too, Sandy Nairne, "The institutionalisation of dissent," in *Thinking about exhibitions*, pp. 387–410.] Interestingly, an early inspirational exhibition that was shown at the Exploratorium was "Cybernetic serendipity," which Jasia Reichardt of the London Institute of Contemporary Art organized in 1967. This was another experimental exhibition that, in this case, embodied the fundamental principle of the observer becoming "a constituent of the work as art"—their presence being the crucial factor that turned the machines into artworks. [See Jasia Reichardt, *Cybernetic serendipity: the computer and the arts* (New York: Frederick A. Praeger, 1968) and Hilde Hein, *The Exploratorium: the museum as laboratory* (Washington and London: Smithsonian Institution Press, 1990), pp. 33–35.]
19. Another example of a sci~art project was entitled "Phantom Limb," which involved the scientists John Kew and Peter Halligan collaborating with the artist Alexa Wright on a project that investigated the self-perceptions of patients with limb loss who experience "phantom limbs"—the well-recognized neurological phenomena of sensing that their missing limb is still there. Wright's digitally manipulated photographs recorded patients' changing senses of their own phantom limbs, and in the process produced some remarkable pictures as well as genuinely revealing new insights into the understanding of this phenomenon, particularly the patients' own perspectives on it. [For further information, visit www.wellcome.ac.uk.]

Virtual Museums: The Challenges of New Technology

New Technologies and the Objects of Science

REFLECTIONS ON THE USE OF MULTIMEDIA*

Paolo Galluzzi

The approach of the passage to the third millenium has stimulated renewed reflection on museums, on their functions in contemporary society and on their future prospects. The discussion has concentrated on museums of art, which have always constituted the privileged space of museology, while the problems relating to technical and scientific museums have remained at the margins. Nonetheless, the complex and variegated world of scientific museums is displaying evolutionary tendencies that merit attention.

My reflections will concentrate mainly on the recent evolution of technical and scientific museums of a predominantly historical character.

As is well known, the fundamental motivations that, from the second half of the eighteenth century onward, nourished the foundation of scientific, technical and industrial museums were, above all, the will to enhance the function of science as a motor for moral, civil and intellectual progress and a guiding light of prosperity, the goal of creating structures provided with the equipment and the competences necessary to train qualified technical operators, and, last but not least, the need of promoting national identity through the celebration of industrial achievements and the ingenuity of inventors.

With time, these motivations became progressively attenuated, or found more appropriate expression in spaces or through channels other than scientific museums. The result was an identity crisis for museums of the history of science, a crisis that brought with it the loss of meaning of the collections and their fossilization in containers often lacking precise goals and effective strategies of communication. Thus, many of the repositories of technical and scientific memory have long remained excluded from the circuit of cultural communication, while showing a scarce capacity to carry out significant exchanges with the core scientific community. The latter has generally seen historical museums as cemeteries of knowledge rather than as resources to be used to present the complex historical evolution of science to a larger public.

The space of communication abandoned by the scientific collections of historical character has been partly occupied by new, radically different structures, such as the science centers. These structures, generally devoid of historical collections, present purpose-built exhibits (that is, not directly derived from the devices used in research), designed on the basis of a "bottom-up" philosophy. Rather than passively receiving information, the visitor himself engenders the process that leads to

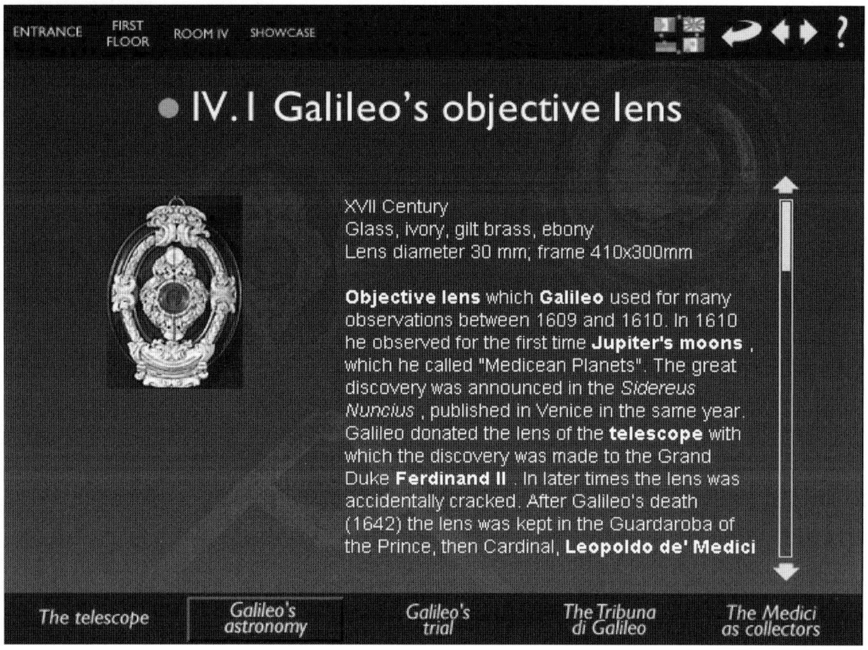

FIGURE 1. The object interface of the IMSS Informative System: Galileo's telescope objective lens.

the observation, the measurement, or the discovery. This approach is sharply distinct from the "top-down" presentation of traditional museums. In the science centers, which have progressively spread throughout the globe, the message transmitted appears inverted with regard to the traditional museum. While in the latter the imperative is "look, but do not touch!", in the former one is explicitly invited to manipulate and explore.

In the last few years, the extremely schematic scenario that I have sketched has shown dynamic processes of great interest.

With regard to the science centers, one observes a vivacious discussion about the need to recalibrate their mission and their strategies of communication.[1] Some experts complain about the risk of a progressive loss of autonomy as a consequence of the fact that the extremely high costs of maintaining and continually updating this kind of structure require that progressively larger exhibition spaces be devoted to promoting the products of sponsoring industries. In addition, science centers tend to privilege the most fanciful technical applications, paying little attention to the presentation of the problematics of fundamental research.

The effectiveness of the imperative of interactivity, with the pedagogical philosophy upon which it depends and with the operational instrument—the "hands-on exhibit"—that distinctively characterizes it, is also being called into question. It has been stressed that "hands-on" does not necessarily also imply "mind-on." Others admonish that the concept of "edutainment," drawn upon increasingly by science centers, urged by the competition with commercial structures that use interactivity for purely ludic ends, risks transforming these institutions into amusement parks.

Finally, observers have pointed to the repetitiveness of the contents and strategies of communication of the majority of science centers, to such a degree that it has been said that having visited two or three of them is to have visited them all. While this judgment is obviously too severe, it does capture a trend of global homogenization which seems to be effectively at work.

If this is the current state of the discussion on science centers, in recent years, there have been encouraging signs of a renewal of interest in scientific museums of historical character and, more generally, in the scientific heritage.

A decisive stimulus to this revival has been offered by the development and continuous refinement of studies of the history of science in the last few decades. In particular, the most recent historiography of science has come to focus ever more intensely on instruments of science and scientific, technical and industrial heritage. Exemplary studies have been devoted to the procedures and devices used in experimental activities, providing effective demonstrations of their complex character, of the interactions between interpretations of experiments, the hardware available and the theoretical framework of reference. At the same time, spectacular performances of experimental archeology have been promoted. Moreover, the influence of social factors in experimental activity has been convincingly stressed. Contextually, thorough studies have been devoted to instrument makers, scientific laboratories and the theatricalization of experiments, while numerous catalogues of little-known, or even altogether unknown scientific collections have been published. Finally, the phenomenon of scientific collectionism has been investigated with new methods and perspectives, displaying its complexity, local variations and function as a site of encounter between demands of promotion and social visibility and utopian enterprises aiming at radical reform of knowledge and of the traditional organization of culture.

Additionally, the attention of historians of science has ceased to focus only on great personalities and revolutionary discoveries, broadening its gaze to encompass both the scientific movement in its entirety and local developments, even those long-considered marginal, contributing in this way to reawaken interest in collections, documents and artifacts traditionally neglected.

Finally, numerous works have demonstrated the contribution which the study of scientific instruments can offer to improve our comprehension of significant aspects of cultural activities which we now regard as completely separate from science. This is particularly evident in the case of the artistic production of the Renaissance. As has been convincingly demonstrated by several studies, Renaissance artists systematically availed themselves of scientific methods and instruments in the preparation and execution of their works. Some of these artists, additionally, invented or perfected important optical and measuring instruments, which are now preserved in museums of the history of science.

This complex series of factors has contributed to bringing about renewed attention to the scientific and technical heritage and to stimulate initiatives that promote scientific museums and collections and make them accessible to a large public. However, this last objective encounters very serious difficulties for a number of reasons.

Above all, the almost complete absence of historical presentations of the sciences and techniques in the formative curricula of schools world-wide implies that visitors to scientific museums bear a handicap of basic knowledge that is normally higher than that which they have in museums of art. The situation does not appear more advantageous for the visitor who has carried out scientific

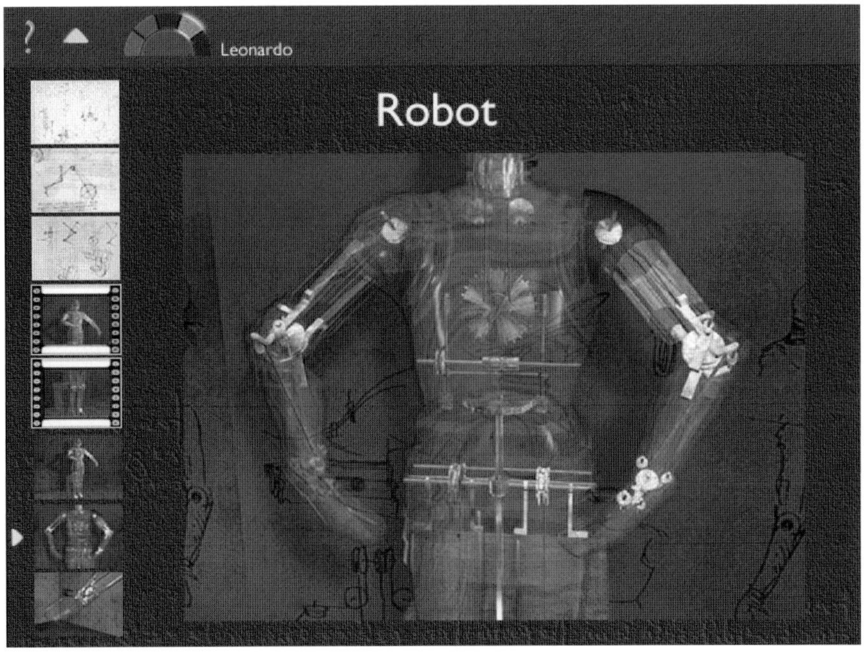

FIGURE 2. Virtual working model of Leonardo's robot.

FIGURE 3. Virtual working model of Vitruvius' odometer.

studies, given that these generally avoid the historical perspective entirely. It follows that many of the material testimonies conserved in this kind of museum remain not only incomprehensible for the scientist-visitor, but often produce embarrassment, and even, occasionally, provoke an instinctive attitude of refusal.

It should also be added that the importance of historical collections of science and technology cannot be fully grasped by means of the simple inspection or even the manipulation of single objects. It is necessary to catch the network of logical, chronological and operational relations which the piece examined bears to other objects (internal and external to the collection), to the conceptual problems of reference and to the standards of analysis, measurement and certification employed in the context in which the object was conceived and used.

A further difficulty is constituted by the fact that this type of artifact can be explained only in an inadequate manner by purely verbal descriptions. For its full comprehension, especially on the part of the non-specialized public, it is necessary to display the dynamics of its operation, and visualize the context in which it was used. This kind of information is sometimes provided by museums by means of graphic schemes, which offer only a partial solution to the problem.

It is obviously not easy to give an effective response to these demands for information with captions and panels, that is to say, with the traditional communicative weaponry of the museum.

The awareness of these limits has stimulated many historical museums to elaborate new instruments and new strategies of communication, in order to restore to the collections they preserve a more effective capacity to capture public interest.

In some cases, an attempt has been made to transfer to historical museums methods of presentation developed in the science centers, that is, by constructing replicas of the original instruments and experimental devices that can be manipulated freely by the visitors.[2]

Further, museums have become more and more involved in initiatives of diffusion of scientific culture and of public understanding of science. There are social responsibilities, rightly considered crucial today, that have not, until recently, been considered as lying within the scope of scientific and technological museums of a historical character.

In Italy, for example, the efforts carried out in this direction have contributed to spreading the conviction that the promotion of historical museums of science and technology forms an essential part of the project to favor a better understanding of science among the citizens.[3] It is evident that the restitution of a social value to museums of the history of science implies a significant expansion of the responsibilities of curators, with respect to the traditional roles of conservation, cataloguing and the exhibition of patrimony.

Another route which has been explored with very significant results in recent years is the preparation of special exhibitions and didactic laboratories by historical museums. The organization of exhibitions needs to be particularly emphasized, given that well-conceived exhibitions often propose a radical reconception of the methods and strategies of communication used traditionally in museums. Exhibitions bring about the recombination of items which have been separated by the historical processes of specialization from which the contemporary system of museums, with their clear distinction according to discipline and genre, have derived. Exhibitions generally reunite instruments and experimental apparatus conserved in different collections, replacing them in the context of the scientific debate of which they were the occasion and/or the expression. They integrate them with displays of books, manuscripts, paintings, drawings, furnishings, costumes, medals, portraits of protagonists, suitably designed exhibits and, above all, reconstructions of the cultural, historical and social background. In this way, the visitor is placed in the best position to appreciate the complexity and multipolarity of the processes of investigation and discovery (including, of course, those "discoveries" which turned out to be erroneous).

Good exhibitions, additionally, also provide the larger public with an effective synthesis of the conclusions reached by the most advanced contemporary research, and thus fulfil the extremely important function of updating information, a function that museums, with their dated exhibits, can hardly ever perform satisfactorily.

In addition to exhibitions, in recent years, in efforts to valorize collections relating to the history of science, growing use has been made of information technologies. In spite of this, I feel that the enormous potential of these new instruments has not yet been fully exploited.[4] The current lively discussion surrounding the presumed risks relating to purely "virtual" consumption of cultural heritage is, in reality, symptomatic of the resistance on the part of some museum curators to use these new instruments. In my opinion, the question has been incorrectly posed. The problem is not that of contrasting the "real" to the "virtual," but rather of using the "virtual" to shed more light on the real object that, if regarded as an isolated atom, risks remaining indecipherable. The dangers do not reside in virtual consumption, then, but in the improper use that is commonly made of this new language in many museums of art, science and technology.

FIGURE 4. Virtual working model of Francesco di Giorgio's mud extractor.

There is, in a word, the risk that the inertia of the traditional function of the museum will lead to a reaffirmation in the virtual domain of the bonds of specialization of collections according to genre and discipline, making it thus impossible to seize the extraordinary opportunity of reviving the cultural and educational functions of the museum offered by new technology.

In fact, these instruments could be used effectively to overcome many of the limits of the traditional strategies of communication of scientific museums described above. They can also produce (particularly in the case of telematic networks) favorable conditions for the establishment of essential collaborations between different museums and between museums and the centers of scientific research, favoring the active participation of museums in initiatives for the diffusion of scientific culture and for the public understanding of science. Conversely, we observe that often the doors are opened to digital innovation, but only to perpetuate in more scintillating, modern and suggestive forms the sacerdotal mission of the museum-temple and the museum microcosm which no longer corresponds to the expectations and demands of today's society.

The idea of the virtual museum which seems to be asserting itself, is, in fact, the idea of the digital clone of the real museum, accepting all of its structural features. In the digital replicas which are produced, the museum's architectonic structure and rooms are reproduced with faithful realism, the lacunae in the collection are accepted without worrying about how they might be filled, and, finally, its internal division into special departments is strictly reproduced. To have meaning and utility, the "virtual" museum should rather be constructed with a significantly different configuration from the "real" museum. In cyberspace, the real museum should constitute only the point of departure and return of journeys that are confined neither to the internal perimeter of the museum, nor to the sole discipline or genre to which it relates. The visitor should be permitted to explore not only objects, but also ideas and persons, places and events; to select different levels of explanation; and to have access to didactic tools.

The nineteenth-century concept of the "museum-temple," a symbolic and evocative site *par excellence*, cannot, then, be transferred as it stands into cyberspace, where there exist neither exhibition rooms nor objects to venerate in the magical "aura" of the museum, capable on its own of evoking emotions that satisfy the soul. In cyberspace, what one encounters are simply images to be observed and decoded with the help of information, relations and comparisons.

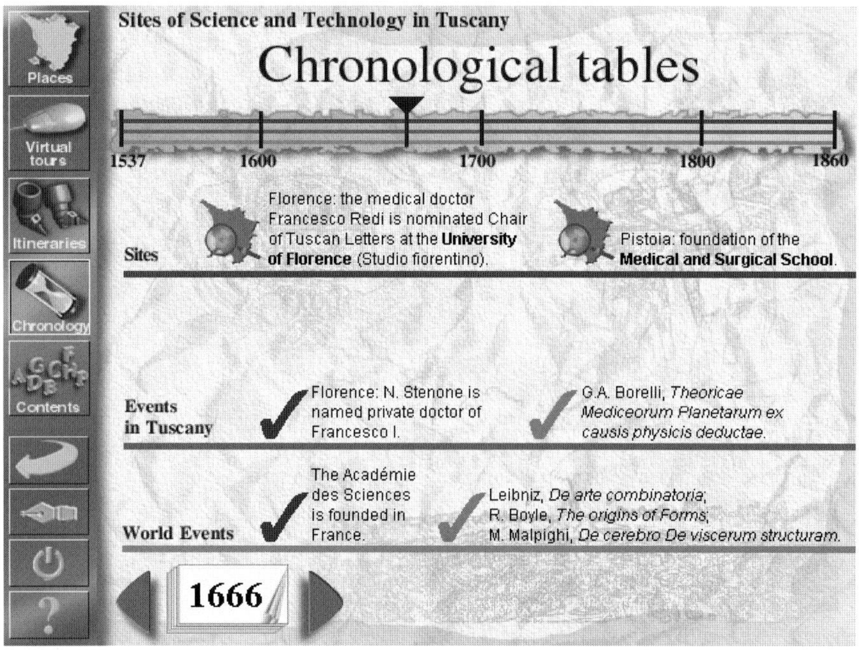

FIGURE 5. Time-line of the hypertext "Science and technology in Tuscany".

In the last decade or so, the reflections and analyses which I have described have motivated the revision of the goals and communication strategies of the Istituto e Museo di Storia della Scienza in Florence. Together with a series of structural interventions, the deployment of new technologies has played a crucial role in these efforts.

I am fully aware that the Florentine experience does not have an exemplary character, nor can it be proposed as a model. Nonetheless, given the significant dimensions of the investment that has been made over a period of more than ten years in human resources and equipment, it may be useful to attempt a first assessment.

The continually increased introduction of information technology into our institute, beginning at the end of the 1980s, has been accompanied by the progressive and non-linear definition of the objectives to be achieved with the aid of such instruments. One might say that we have endeavored to overcome, with these tools, the limits of communication of historical collections in the scientific and technical domains which I have discussed previously, limits which have long excluded scientific museums from the circuit of cultural communication. From the end of 1994, moreover, our institute has launched its Web site, conceived as a focal point for the expansion of our communications strategies and as a service center for the enormous and diverse group of users (researchers, students and private individuals) potentially interested in the material we have to offer, on a planetary scale. In particular, the Web site is currently configured as an information desk[5] continually updated on scientific culture, a place for the promotion of the initiatives of our institute, and a space to collect the requests, expectations and evaluations of those who consult it, thus helping us to improve the service continuously. The remote user is a new reality, an important reality, and merits the greatest attention even if he or she does not pay the entrance fee to the museum.

The concise demonstration of our production that I will submit to you includes both on-line and off-line products, and presents both completed projects and work in progress.

I would like to bring your attention, in the first place, to the statistical data relating to the public audience of our Web site. These indicate the truly extraordinary impact of communication via the Internet, allowing us a direct testimony of the enormous responsibility for the certification of the quality of information which falls upon curators of museums that are active in cyberspace.

I will not elaborate in this venue on the research projects (Ms. 72, Kircher correspondence, Mss. of the Florence National Library, Galileo's integrated Archive), or on the tools constructed to offer qualified services to the international community of historians of science (such as special bibliographies, indices of historical journals, and so on).

With regard to museum collections, the model that we have developed, available both on- and off-line (a multimedia informative system which covers all the items on display and which will be accessible to visitors thanks to computers located in the exhibition halls), has been conceived to respond to three fundamental objectives:

1. To allow the visitor virtual access to the displayed objects, permitting their localization in the physical space of the museum. This may be regarded as the first level, that of the virtual exploration of the Museum collection [Quick Time].
2. To integrate this first level with information and possibilities of further exploration that respond to some fundamental requirements
   a. To stimulate a full understanding of individual objects offering a fully explorable photographic reproduction and a detailed hypertextual description.
   b. To enhance the relationships between the objects and their context of reference. This goal is achieved by offering the user the possibility of consulting a series of thematic slide shows which provide the textual and visual information indispensable to catching the real meaning of the item and of its intellectual and material links with the prevailing historical, cultural and social context.
   c. To favor the visualization of the actual working of the object and a clear understanding of the principles on which this is based using films, computer animations, graphic simulations and so on.
   d. To favor the comparison and integration of the objects of our museum with congruent objects possessed by other institutions. This

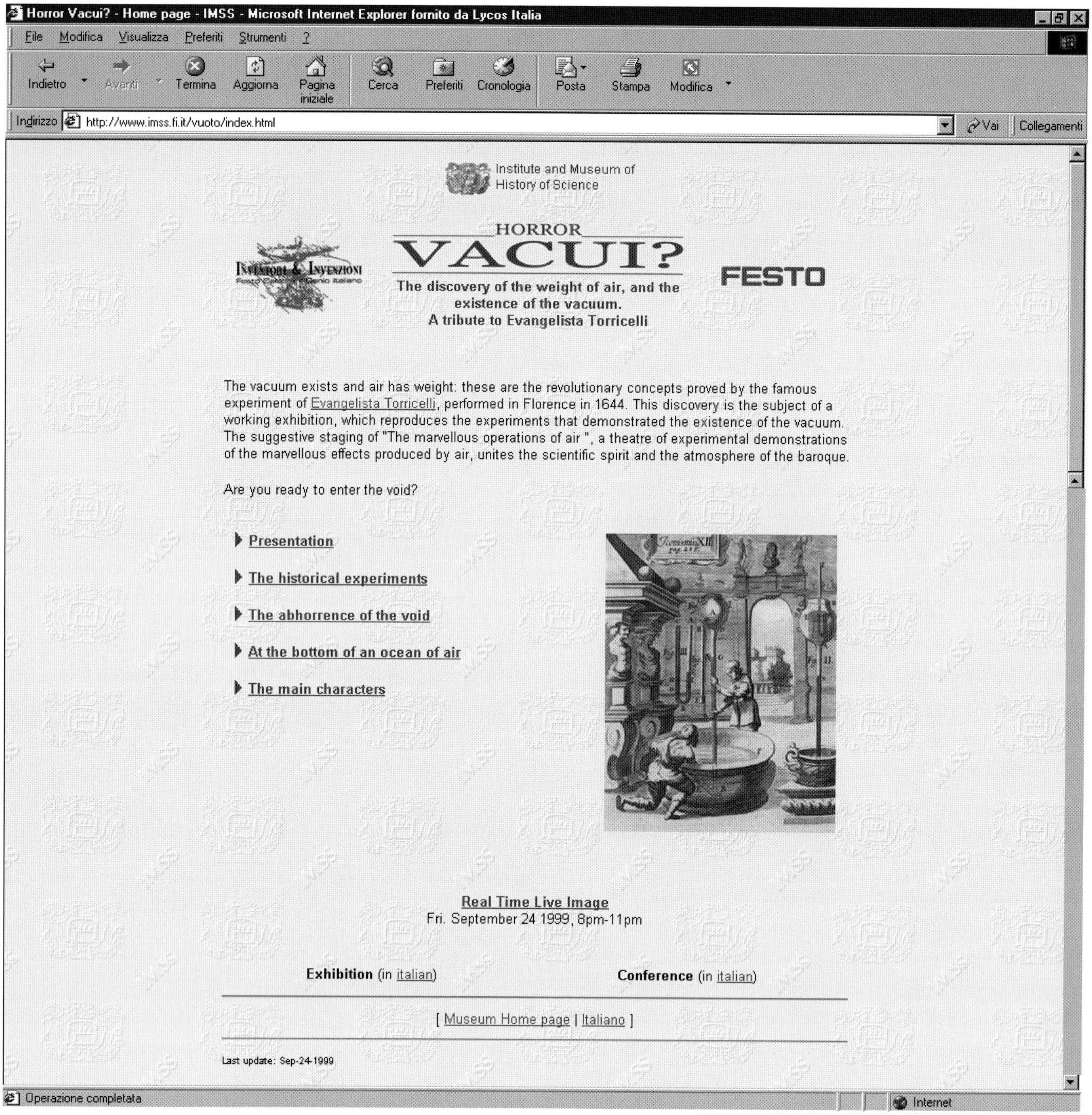

FIGURE 6. A screen shot from IMSS Web site: the virtual exhibition *Horror Vacui*.

is the goal of a project, called "Epact" (European pact), developed in cooperation with the Museum of the History of Science (Oxford), the Museum of the History of Science (Florence), the British Museum (London), and the Museum Boerhaave (Leiden), thanks to the support offered by the European Community.

e. To provide instruments allowing deeper knowledge of general themes (offering the user the possibility of exploring bibliographical and archival resources related to the items examined).
f. To correct the traditional and exclusively "object oriented" approach of visitors to Museum collections. This means giving the

user the possibility of taking specific topics and conceptual themes as a point of departure for the visit to the collection and helping him to isolate the items related to that topic. In this way the visitor can deconstruct the physical display of the objects in the museum and create different strategies of exploration according to his personal interests and curiosity, and/or research targets.

This multimedia application already represents a radical enhancement of traditional strategies of communication, an enhancement that will be further accentuated in the future.

There is a third goal, which we try to achieve through a tool that is already in operation, designed so as to be integratable with the information system illustrated above. This is a hypertext, consultable both on the Internet and off-line, entitled "Science and technology in Tuscany," which allows the visitor to pass from an inspection of the Museum collections to a virtual visit to the places of interest for the history of science and technology in the Tuscan region. Providing an "exploded view" of our collection in its geographical context, this tool stimulates and assists the development of scientific tourism, a phenomenon which is still unfortunately rare.

On the other hand, the temporary exhibitions—both the "real" and the "virtual" ones—promoted by the Istituto serve as important instruments of integration of the products described above. These are organized so as to render visible—by means of multimedia, 3D-reconstructions and animations—the most striking evidence of the decisive role of technical and scientific knowledge in the construction and development of all civilization, as shown through applications devoted to the two emblematic cases of Pompei[6] and Renaissance Florence.[7] Due to the emphasis traditionally laid almost exclusively on artistic and architectural features, the key role played by science and technology remains almost invisible to the millions of visitors who flock to these centers.

This set of integrated applications has elicited an extremely positive reaction, both on the part of the public visiting the museum and the exhibitions, and on the part of the far larger number of visitors that consults the Web site.

The effort that can be observed on an international scale regarding experimentation with new methods of employment of new technologies offers the possibility of continuous refinement of the best methods and models to deploy these tools to promote the technical and scientific heritage.

The scenario which has been opened up by digital technologies thus appears to represent a sort of fourth dimension of communication, by which scientific museums can profit, especially if the use of these new tools is guided by a clear project and a mature definition of the objectives to be achieved.

NOTES

*Translation from the Italian by Michael J. Gorman.

1. For the major topics under discussion, see *Here and Now. Contemporary science and technology museums and science centres*, eds. G. Farmelo and J. Carding. Proceedings of a conference held at the Science Museum, London, 21–23 November 1996 (London, 1997).
2. An excellent example of this strategy is offered by the interactive exhibits displayed in the Boerhaave Museum in Leiden alongside the original instruments.
3. See my paper "Un project strategique de culture scientifique," "Alliage" [*Science et culture en Europe*. Numero special], 16–17 (1993), pp. 101–108.
4. I have discussed this potential more at length in *I formati della memoria. Beni culturali e nuove tecnologie alle soglie del Terzo Millennio*, eds. P. Galluzzi and P. Valentino (Florence: Giunti, 1997). See, in particular, P. Galluzzi, "Nuove tecnologie e funzione culturale dei musei," ibid., pp. 3–39.
5. Currently (October 1999), the Institute and Museum of the History of Science Web site (*http://galileo.imss.fi.it/*) hosts more than 20,000 html pages.
6. The exhibition *Homo Faber. Scienza e Tecnologia nell'antica Pompei* was organized by the Archaeological Superintendence of Pompei and by the Institute and Museum of the History of Science in Florence. The first installation was inaugurated in Naples on 25 April 1999 [see catalogue of the exhibition *Homo Faber. Natura, scienza e tecnica nell'antica Pompei*, a cura di Annamaria Ciarallo, Ernesto De Carolis (Milan: Electa, 1999)]. In October 1999, the exhibition was opened at the County Museum, Los Angeles. The multimedia laboratory of the Institute and Museum of the History of Science in Florence created the CD-ROM of *Homo Faber* in an English and an Italian edition. The CD-ROM is distributed by Elemond, Milan.

7. The exhibition *Gli ingegneri del Rinascimento. Da Brunelleschi a Leonardo* [catalogue in Italian and in English: *Renaissance engineers. From Brunelleschi to Leonardo da Vinci*, ed. P. Galluzzi (Florence: Giunti, 1996)] was created by the Institute and Museum of the History of Science with the fundamental financial support of FINMECCANICA. The exhibition was inaugurated in Paris in November 1995. It was then on display in Florence (April 1996–February 1997), in New York (October 1997–March 1998), and in London (October 1999–April 2000). The CD-ROM, created by the multimedia laboratory of the Institute and Museum of the History of Science, and distributed by Giunti, Florence, was presented in October 1997.

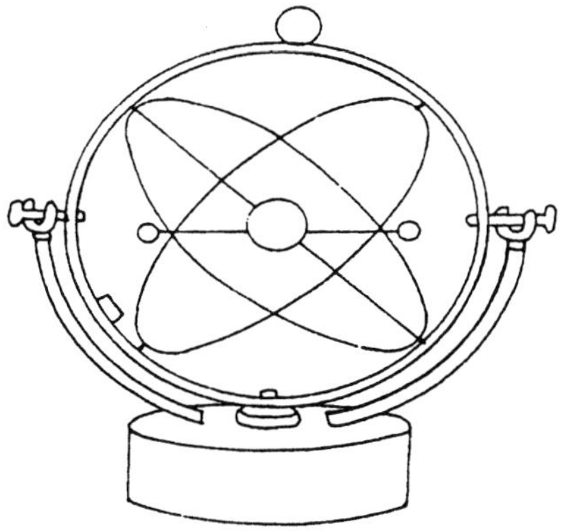

From Hands-On to Minds-On

CREATIVITY IN SCIENCE MUSEUMS

Saroj Ghose

In one of the ancient Indian philosophical and theological treatises, the *Soul* is described as an immaterial entity which cannot be cut by weapons, nor burned by fire, nor be wet by water, nor immersed in air. In Hindu belief, the Soul is ageless, immortal, and indestructible. Death and reincarnation of life constitute an eternal cycle wherein the *Soul* leaves a body much like one deserts a worn-out cloth and on rebirth takes up a new body like a new cloth.[1] The existence of Soul is still an unsolved mystery and so is the *Mind.* But there is a difference. The notion of mind as an immaterial substance akin to the Soul, as conceived by some of the ancient schools of philosophy or theology, is fast losing ground to the current concepts of psychology which view mind as a physical part of the body, encompassing the brain and the central nervous system, and involving physiological processes open to study by biologists and chemists.[2] This study takes note of a more systematic view of mental function that includes perception through sensory organs, assimilation through learning, retention in memory, information processing through thinking, and motivation for action as inseparable aspects of the mind. Although the subject is still not fully understood, our perception of the material world through sensory organs has been a matter of curiosity since time immemorial.

THE WORLD OF PERCEPTION

In the 2,500-year-old *Samkhya-Patanjala* philosophy of ancient India, the world is composed of *Triguna*, i.e. three virtues—namely, *Sattva*, *Rajas* and *Tamas*, signifying the intelligence-stuff, the energy-stuff and the matter-stuff.[3] The three *guna*, though assuming infinite diversity of forms and powers, can neither be created nor destroyed. The totality of the energy-stuff and the matter-stuff remains constant, representing together the manifested and the unmanifested, the actual and the potential. It is only the changes or differences in the aggregation which are responsible for the diversity of the phenomena of objects.

The same philosophy developed a concept of *Tanmatra*, with five subtle infra-atomic particles imperceptible to human senses, but still linked with the five perceptions of *Sabda* (sound), *Sparsa* (touch), *Rupa* (vision), *Rasa* (taste) and *Gandha* (smell). The five *Tanmatra* give rise to five grosser elements, *Pancabhuta*, consisting of *Akasa* (space), *Vayu* (air), *Tejas* (fire), *Ap* (water) and *Ksiti* (earth).[4] The concept of five basic elements in Indian philosophy is akin to the almost contemporary concept of four elements in the Greek philosophy, with an additional element of space. What is interesting in the

Indian concept is the close link between the attributes of perception through five major sensory organs and the attributes of five basic elements that constitute the universe.

This link between the material world and perception, so far an inexplicable mental function, is vitally important. Our senses, through five sensory organs, assimilate information about the external world surrounding us and this helps us in the internal processing of information in the brain. Audition collects information about the remote and unseen, tactile perception brings in a real hands-on situation, vision helps us in locating our bearing, and taste and smell give a personal feeling about our environment. Our senses collect material information that passes on to the brain for internal analysis and interpretation. Certain not-so-well-defined mental functions—viz., ideas, imagination, inspiration, emotions, etc.—get involved in this process. Until very recently, consciousness and the physiological basis of emotions were not considered valid topics for scientific study. Now these topics are rapidly becoming an acceptable and even urgent subject for scientific investigation. Unlike the state of knowledge of the physical universe that has been acquired through the natural sciences, the state of knowledge of the overall workings of the mind is still in its infancy, yet it looks probable that internal analysis in the brain would require adequate input information through all possible sensory organs.[5] The teaching-learning process is no longer simply audio or visual; nowadays it summons all kinds of perceptions and participation through trial and error. A modern writer on brain function rightly observes that "the difference between amoebae and Einstein is that, although both make use of the method of trial and error or elimination, the amoebae dislikes erring while Einstein is intrigued by it; he consciously searches for his errors in the hope of learning by their discovery and elimination."[6]

THE NEW CONCEPT OF MUSEUMS

This concept of learning through a discovery process, with the help of hands-on or minds-on exhibits and activities, led to a series of changes in the concept of museums in recent years. The first change in the concept of museums came through thematic presentation in the galleries. Traditionally, museums were developed in subject-based water-tight compartments—sculptures, paintings, bronze, decorative art, manuscripts, numismatics, etc. Natural history museums were developed into separate galleries on invertebrates, insects, reptiles, mammals, primates and so on. Science museums had subject-specific galleries on electricity, optics, electronics, communication, space, medicine, transport, energy, etc. The traditional approach was academic and, in fact, a bit bookish. In recent times, natural history museums have come up with new galleries, not on academic subjects but on multi-disciplinary themes like *life in the ocean*, *ecology of deserts*, *wealth of forests*, etc. For thematic presentation in museums other than on natural history, I would like to cite three specific examples of the new approach. A new gallery on *hair styles* was developed in Haifa, Israel, with exhibits picked up from the collections of sculptures, seals, coins, paintings, murals and bronze. Here, the stress was not on the objects *per se*, but on the particular theme, viz., evolution of hair styles over the ages. The second example is the series of new museums of civilization in Canada and in eco museums in different countries. In museums of this kind, the center stage is occupied not by the objects but by the contextual significance of objects that build up a storytelling museum. The third example is the Holocaust Museum in Washington, D.C., where storytelling is complete with objects, graphics and video. It is like fluently reading a complete chapter of history, without any gap between the objects.[7] This kind of *total* presentation in new galleries or museums has introduced a minds-on concept, although the exhibits are not necessarily hands-on.

Looking at the science museums, we notice that since the days of the Crystal Palace Exhibition of 1851, science museums have come a long way from being repositories of artifacts to wonderlands of science. On the one hand, the concept of a collection-based science museum changed slowly but steadily to period settings and thematic presentation, as was successfully done in the National Museum of American History in Washington, D.C., and on the other hand, science museums started answering the *what*, *why* and *how* of science through carefully designed exhibits to explain scientific principles or demonstrate functions of a contrivance or project multiple solutions of a problem through massive visitors' participation. Understanding of science through personal involvement and enjoyment became the key design criteria for such exhibits. Side by side with exhibit-based science museums arose a new type of action-oriented science center with interactive exhibits and futuristic presentation. The new trend introduced a qualitative change in museum curatorship in such science centers. Scholastic research on collections yielded to creative design of hands-on exhibits, and historians were replaced by scientists and engineers in these kinds of science centers.

The new wave of ideology in science centers understandably brought in a brand new generation of exhibits. Visitors were encouraged to press a lever, turn a wheel, roll a ball, pull a string, look through a peephole, and be psychologically involved with animated exhibits. The new class of exhibits was termed *participatory* because of the massive participation of visitors, but like many good things, the term soon became a *cliché* through repetitive usage without context. Some museum curators argue that in effect every exhibit having a visual or tactile appeal becomes participatory the moment the visitors absorb information by looking at them or by touching them. Some believe that all push button exhibits are participatory, as they require manipulation by visitors. Others contest the idea by saying that only exhibits that involve active motor action of the brain can truly be called participatory. For a better understanding of the definition of participatory exhibits, its nexus with *interaction* has to be clearly established.

INTERACTIVE EXHIBITS

Interaction, and in contrast non-interaction, depend on the mode of communication between the two parties involved in a program. In my opinion, interaction should have at least one of the following two criteria:

(a) two-way communication
(b) multiplicity of options

Passive exhibits are those which have visual or tactile appeal but are non-working in nature. Active exhibits are those which are animated or work in the same manner every time they are operated. Interactive exhibits are those which build up a dialogue with the visitors by operating in different manners and by throwing questions to visitors through such operations. These are generally hands-on exhibits. Minds-on exhibits, which *may or may not* be hands-on, are those exhibits where visitors go through a discovery process by themselves or cross the barriers of time and space in a simulated situation. These are shown in Figure 1 and explained in subsequent paragraphs.

FIGURE 1. Passive, active and interactive exhibits—Hands-on and Minds-on.

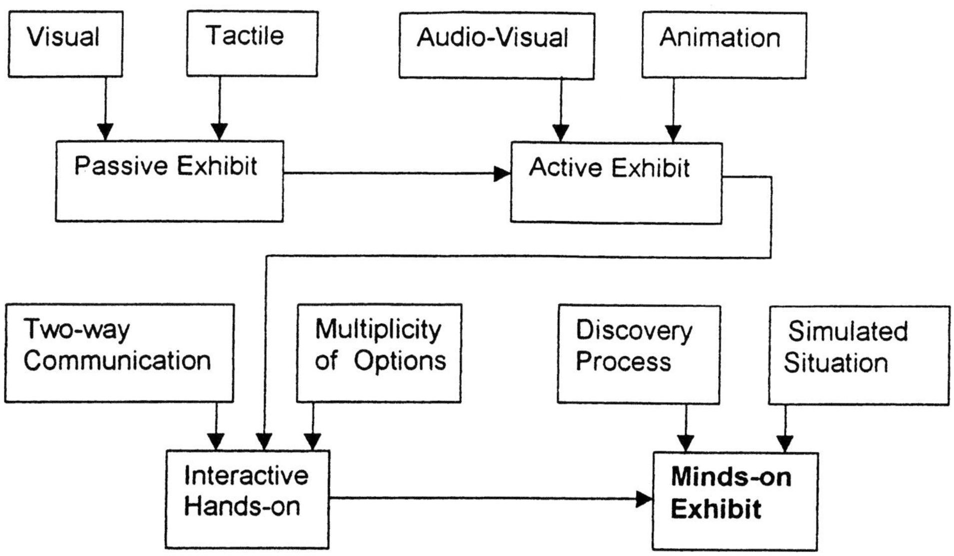

The best means of providing two-way communication in a science museum is evidently a computer or microprocessor circuit, which involves visitors in a quiz or matching game or storytelling or problem-solving activity through a back-and-forth transaction of information. Computers can be programmed to behave almost like human beings by answering questions, throwing counter questions, confronting visitors with problems, and grading the performance of the visitors. Many computers have appropriately designed software, involving interactive color graphics. Personal computers are becoming very useful tools for science communication. Interesting variations of this class of exhibits are interactive video programs or computer multi-media. With touch screen sensors or a mouse or a track ball, visitors can retrieve any visual—still or movie—at random from the visual information bank and follow a story or play a challenging game with the computer. Much simpler and economic versions are microprocessor-controlled quiz or matching games, with the exhibits throwing questions at visitors and grading their answers. Much more complicated is the world of virtual reality, wherein hands-on concepts are transcending into minds-on perception.

We can have a look at some simple mechanics exhibits wherein two-way communication or multiplicity of option is not so explicit as in a computer-aided game but imbibed in the operation. These exhibits interact not with spoken or written words or graphic expressions like computer multi-media or interactive video programs but interact with their behavioral pattern. An exhibit on coupled oscillation is a simple example (Figure 2). Visitors set one pendulum bob to swinging while the other is at rest. The oscillation is gradually transferred to the second bob while the first bob slows down. It stops altogether when the second bob swings with its maximum amplitude, and then the oscillation gets transferred again to the first bob. The cycle goes on and on. Visitors like to operate the exhibit by pulling the bob in different directions and with varying magnitudes and observe its behavior.

A second example is known as Galton Board (Figure 3). By turning the board upside down, small ball bearing balls are made to fall on evenly spaced pegs and collect at different chambers below, after random deflections from the pegs. It is seen that the balls collect mostly at the central chamber and taper off toward both sides according to a Gaussian curve. But a strange thing happens when a viscous liquid is used in the same exhibit. If the board is turned fast, the liquid collects only at the central chamber without following the Gaussian curve. But if the board is turned slowly, the liquid also follows the Gaussian curve. In both the exhibits on coupled oscillation and liquid Galton Board, the exhibits do not always respond in a pre-

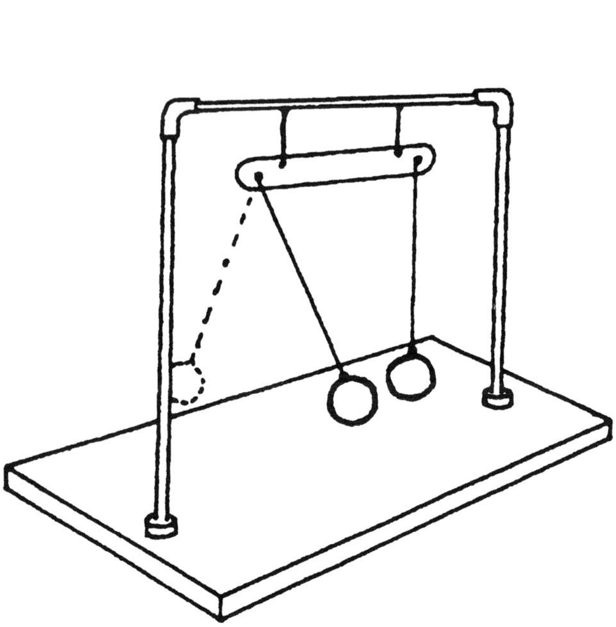

FIGURE 2. Coupled oscillation.

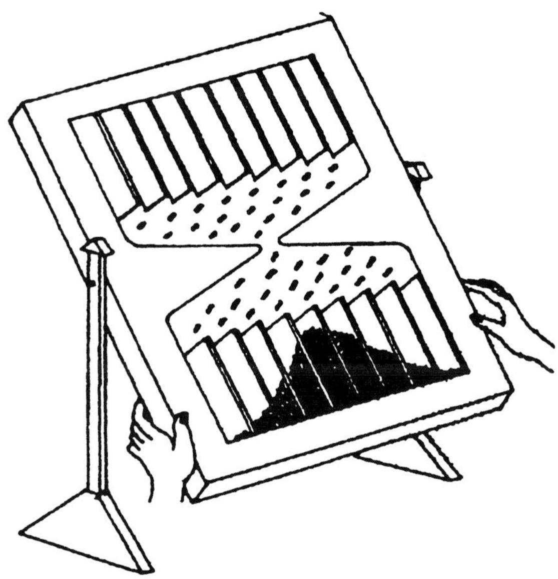

FIGURE 3. Galton board on probability.

determined way. Through a somewhat unpredictable manner, the exhibits throw a puzzle to the visitors to solve and therefore establish a two-way communication. There are numerous examples of such kinds of exhibits presently being developed in science museums or centers.

The second criterion of interaction is the multiplicity of options. If the exhibit operates in a single predetermined manner every time a button is pressed or a lever is pulled, the exhibit no doubt becomes active but cannot be called interactive. For interaction, visitors must have options to operate the exhibit in many different ways and the exhibit must respond in different manners, thus throwing a puzzle or question to the visitor and engaging him in a problem-solving effort. A typical example of this kind of exhibit is a loop through which a rolling ball passes without falling down (Figure 4).

If the ball is released from the top of the channel, it passes through the loop easily. But what happens if the ball is released from a lesser height? The exhibit engages the visitor in an experiment in which he releases the ball from different heights to discover by himself the optimum height below which the ball fails to negotiate the loop. But the experiment does not stop there. The visitor tries to give an additional push or a spin to the ball released from a lower height. He tries to push the ball from the lower end of the loop upwards (Figure 5) to see whether it rises up to the other end after negotiating the loop. Through such attempts, the visitor exercises different options to operate the exhibit.

A second example in this category is shown in Figure 6. Here the visitor releases a hollow empty ball from the top of a curved channel to see how it accelerates down. He then uses a ball half full of sand and finds that it does not roll at all; it slides down. He takes out some quantities of sand and tries his experiment several times to find that an optimum quantity of sand makes the ball just roll, not with acceleration but with constant velocity. This is a simple way of setting a falling object at

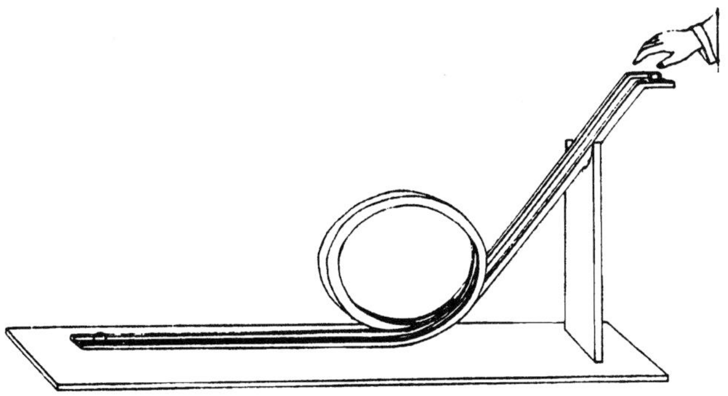

FIGURE 4. **A ball rolling down the loop.**

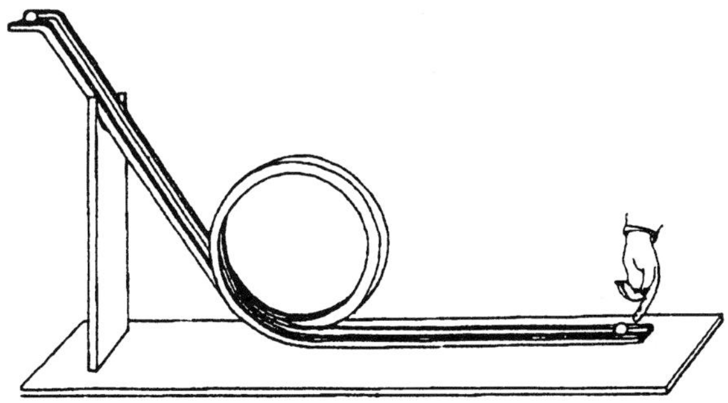

FIGURE 5. **Visitors try to push up the ball.**

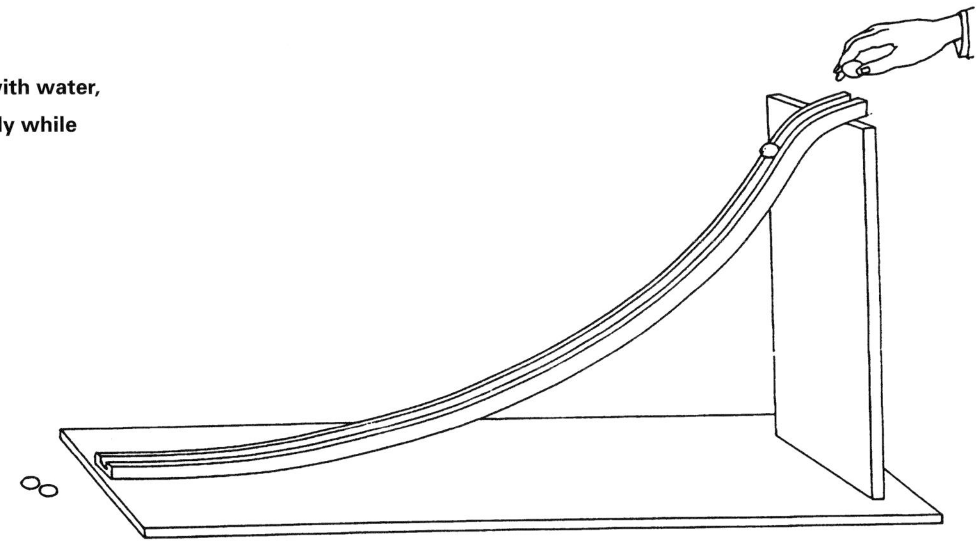

FIGURE 6. **Some balls filled with water, sand and wax behave differently while rolling down the channel.**

constant velocity. If he tries with a ball partially filled with water, he finds that it rolls down exactly like an empty ball and does not slide down like a sand-filled ball. What happens if the ball is partially filled with molten wax and allowed to cool? The ball slides down if it is released with wax at the bottom and hops like a frog if released with wax at the top. The exhibit gives many options to the visitors.

MINDS-ON EXHIBITS

Interactive participation through multiple options of operation and/or two-way communication between an intelligent visitor and inanimate exhibit often lead the visitor to the third criterion of interaction—learning through a discovery process. Through such exhibits, the visitors are urged not to accept a statement given in the descriptive text without verifying the same. The visitors are urged to discover the solution by themselves through repeated experiments that reflect a true interaction and interactive participation. One may be tempted to suggest that the third criterion appears to be a corollary of the first two, but there are examples to show that even in the absence of the first two criteria, learning through a discovery process can make an exhibit sufficiently interactive, although not participatory. If one looks at the exhibit of jumping bobbin (Figure 7), one finds that due to the action of eddy current the bobbin continuously shuttles back and forth between two plates. If he looks at the exhibit shown in Figure 8, he finds that swinging bobs sometimes oscillate and sometimes turn over in a totally unpredictable manner. In both exhibits there is no switch to press, no handle to crank, no physical participation of visitors. Yet visitors' attention is arrested by apparent perpetual motion that lures them to explore from different angles the hidden driving force for the motion. Such exhibits induce the visitors to a lively discussion among themselves in an attempt to discover the real cause after they are convinced that the effect is neither perpetual nor supernatural. Although not hands-on, these are true minds-on exhibits.

The entire teaching-learning process through the minds-on approach can be demonstrated through certain other exhibits which are designed to confront the visitors with challenging questions. The lazy tube (Figure 9) is one such exhibit. This is a thick-walled aluminum tube through which a piece of wood, or acrylic, or iron, or anything else drops down fast due to gravity, but a small magnet takes a long time to fall, although the tube is not made of magnetic material. In fact, with a carefully designed exhibit, the delay could be one minute in a 5-meter tube. It takes some time to realize that, due to eddy current, the magnet in motion sets up an opposing magnetic field that retards its motion. The phenomenon is explained in an open exhibit where several rings slide down fast along a thick aluminum channel, whereas a ring magnet rolls down very slowly (Figure 10).

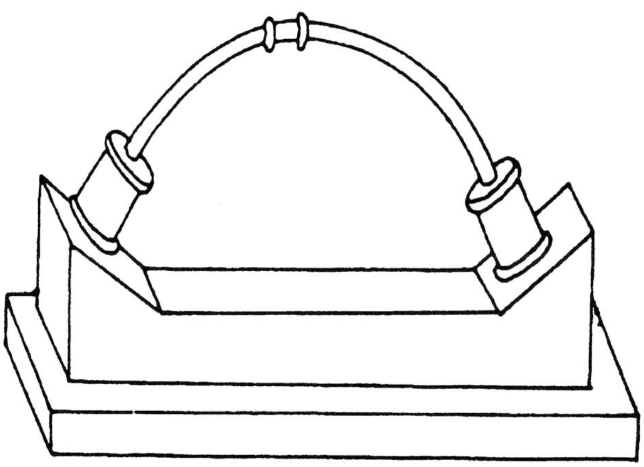

FIGURE 7. A bobbin in perpetual motion?

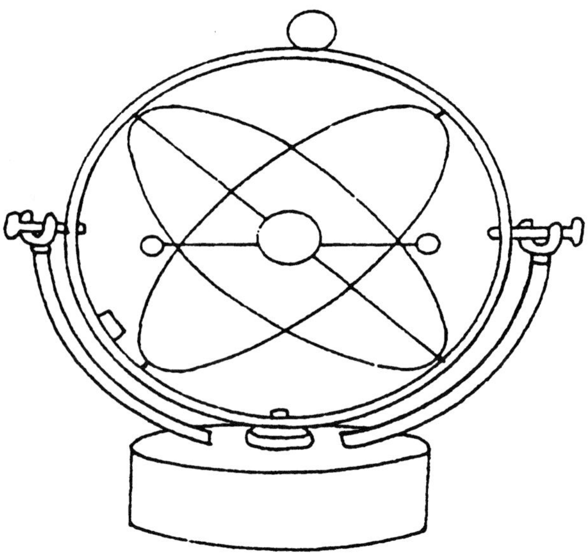

FIGURE 8. Random swing of the bobs.

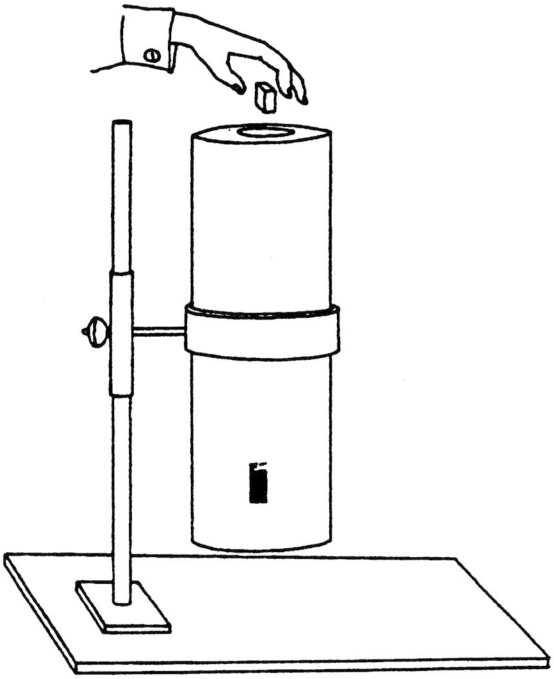

FIGURE 9. The lazy tube—the magnet falls down slowly.

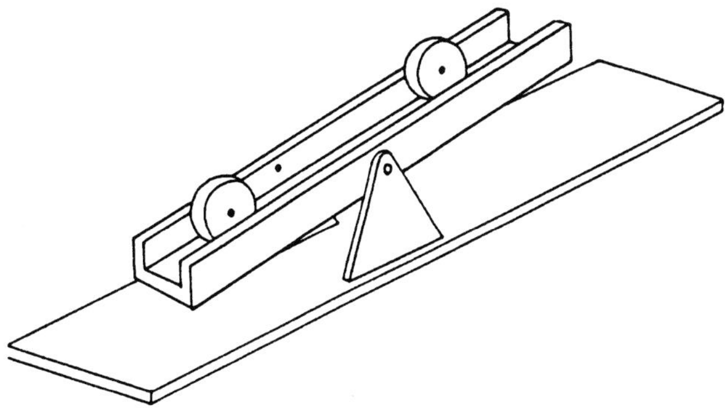

FIGURE 10. The lazy channel—the ring magnet falls down slowly.

The exhibits, cited above, illustrate that it is not the microprocessor jugglery nor the video stunt that alone makes minds-on exhibits. In order to make minds-on exhibits the channel of communication, both in exhibits and activities, has to follow a specific logic pattern. The various steps to follow in such exhibits and activities are:

(i) assimilation of information
(ii) understanding of problem
(iii) formation of hypothesis
(iv) testing the hypothesis, and
(v) reaching an inference

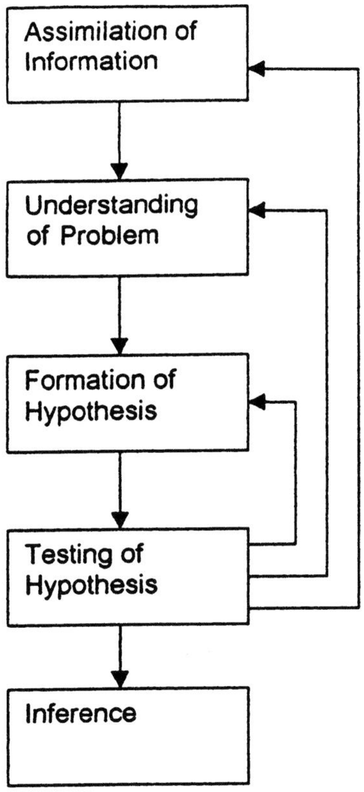

FIGURE 11. Logic pattern for minds-on exhibits.

While testing the hypothesis at stage (iv), a visitor often finds his hypothesis wrong and in such cases he has to come back to stage (iii), form a new hypothesis and go on testing. He may be required to move back and forth several times between stages (iii) and (iv) and sometimes is required to move back to stages (ii) and (i) for a better understanding of the problem (Figure 11).[8]

In a minds-on exhibit, he has to understand every question and find out the answer by himself. The most successful minds-on exhibit will be that which does not give an explanation or answer to the problem in the descriptive label, and the visitors are able to find out the answer through a discovery process. Instead of providing a direct answer through descriptive labels, the exhibit leads the audience to discover the answer through a systematic search and not by trial and error. Even if an answer is found in an empirically cognitive manner, a visitor has to understand the steps through which the answer is worked out. The minds-on exhibit must ensure that it is not the answer that is so important, but the process of discovering the answer that leaves an everlasting impression on a visitor's mind and makes the exhibit highly retentive. In other words, it is the method of science and not science alone that characterizes a truly minds-on exhibit. A good example of a minds-on exhibit of this kind is a continuous water tap that hangs freely at the center of a small enclave, held by thin strings. The enclave is dimly lit and a narrow spotlight is focused on the tap to show clearly that it is not connected to any pipe and still water comes out of its mouth continuously (Figure 12). While every other exhibit in the museum is explained by descriptive texts, this is the only exhibit which throws a question to the visitors: *Where do you get this water from?* It is expected that the visitors will find out the answer by themselves.

The exhibit, first installed in the science museum at Calcutta, was always surrounded by a large group of visitors engaged in animated discussions. Some suggested that water comes down through the narrow strings. Some others believed that the spotlight focused on the tap is converted into water. When some people expressed doubt about this hypothesis, the proponents argued that if the science museum people could do miracles in the electronics gallery by converting light into electricity or sound, they could very well convert light into water. Still, nobody was convinced.

The speculations from visitors encouraged us to develop a complete minds-on hall in Bombay, where the water tap is placed at the center with a big question mark and with no answer. Visitors are asked to discover the answer through a series of hands-on exhibits surrounding the centrally placed water tap. The hall starts with the first hypothesis that water comes down through the thin strings. From the first set of hands-on exhibits, the visitors learn about the cross-section of tubes, flow of water, the concept of volume and they soon discard the hypothesis after observing that so much water cannot flow through the narrow strings even if they are hollow. They try the second hypothesis in a series of hands-on exhibits. They convert light into electricity, sound, heat, and mechanical energy and learn that energy can be transformed from one form to the other but not to matter. They form another hypothesis that water is created at the mouth of the tap by combining oxygen and hydrogen. Through some experiments they learn that for combining hydrogen and oxygen certain requirements are to be fulfilled and these do not exist at the mouth of the tap. While testing each hypothesis, the visitors shuttle back and forth between the water tap and the explanatory exhibits in the same way as is indicated in Figure 11. At some point in time a hint is given: *Do you hear a sound?* Some visitors soon find out that the sound is that of a pump. They face the next question: *What does it do?* An interactive quiz tells them that pumps are used for lifting water up. Well, then there is a pump hidden somewhere, and what else is hidden? In the next set of exhibits, they learn about refractive index. They immerse several pieces of glass in water and find that one glass tube, having the same refractive index as that of water, is invisible in water. After intense group discussions, they infer that there must be a glass tube fixed at the mouth of the tap and water is pumped up through this tube by a pump hidden below the table. Since the tap is closed, water falls down, surrounding the tube, which is not visible inside the water column. Finally you got it! This is an elaborate discovery process.

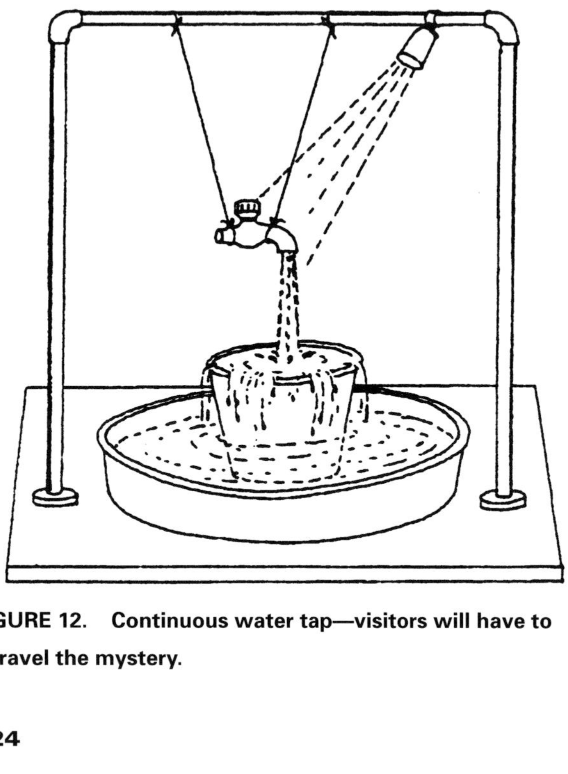

FIGURE 12. Continuous water tap—visitors will have to unravel the mystery.

The second criterion, in addition to *discovery process*, of a minds-on exhibit is the creation of a simulated situation with which the visitors will identify themselves and cross the barrier of time and space for the time being. This is done by faithfully recreating a walk-through period setting with artifacts of the time and sometimes animated with sound and light effects. In a science museum, this sort of simulation could create a dramatic situation in which they could undertake a space flight into outer space or could ride a time machine which takes people to the time of creation of the solar system. In such a minds-on situation, visitors could go back to the age of dinosaurs, with computerized robotic giant reptiles moving in their Jurassic habitat. It could be a fantasy land where the people could travel through the micro-world of atoms. Visitors could fiddle with the molecules that create life and imprint hereditary traits. Exhibits of theme parks could be appropriately modified, with a strong emphasis on science with a view to create a world of virtual reality, with imagination running high but still contained within the domain of real science.

The efficacy of minds-on exhibits largely depends on the imagination and creativity of the exhibit designer. But how do the visitors receive the message? Where is the seat of consciousness within the brain? How can we account for a continuum that begins with cells and neuro-chemicals and ends with such ethereal qualities as imagination and creativity? Modern researchers believe that, through a series of feedback loops, neurons are capable of forming the things that are difficult to fathom, such as ideas, images and the other results produced by the brain.[9] The subject is still speculative.

CREATIVITY IN ACTIVITY

Interactive participation is a creative process not only for the audience but for the exhibit designers and museum curators as well. The success of this process largely depends upon the imagination and creative thinking of the communicator, who could be either the educator or the designer or both. With passive exhibits, the communicator imparts education, formal or informal, serious or through fun. With active exhibits, the communicator induces his audience with investigation, although with no feedback. With hands-on participation, he involves the audience in a scintillating dialogue with his inanimate objects. With a minds-on approach, he gives life to his objects and enlivens the museum as a whole.

Creativity is necessarily not an individual activity nor does it always involve cultivation of the faculties of individual minds. In one way, creativity has a root in the cumulative pool of community knowledge. The debate over the existence of innate ideas vs. the mind as a clean slate still exists in some form within certain schools of thought. One notable example of this debate involves those followers of the theories of Carl Jung who believe in some form of race memory or some form of collective unconscious inheritance from earlier generations.[10] Factors involved in creativity often include collective attention and mental focus, societal beliefs and expectations, cultural traditions and community experiences. Creativity is often demonstrated through collective activity.

For any scientific institution, museum or otherwise, the primary objectives are to foster a spirit of inquiry, to kindle the fire of imagination, and to cultivate creative talent particularly among the younger generation. This objective stands on a foundation of basic understanding of science which can be achieved not only through exhibits but also through well-orchestrated outreach programs involving millions of people every year, an activity in which science museums and centers have specialized over the years. Year round exposure-oriented activities are organized both for the school and for the community. Thousands of school children are involved in discovery rooms, science fairs, environment camps, amateur astronomy, aero-modeling and many other hobbies. Teachers are engaged in development of low-cost teaching aids and science kits. Extensive rural programs are organized that involve people of different spectra of life in science and technology appropriate to their day-to-day living. A science literacy program is gradually gaining ground through science museums and centers. However, if these activities are just transplanted from one place or one country to the other without any ground preparation or soil treatment or natural growth of roots, they will remain as decorative plants in a living room. Activities that are organized on specific demands from society take deep root in the soil and probably help to bring about a social transformation. I would cite some examples.

The first science museum in India moved out of its four walls in 1966, when its first museo-bus, carrying 24 working exhibits on *Transformation of Energy*, rolled out of the museum. This was the beginning of a large fleet of 25 museo-buses traveling through the deep interior of the country at a later stage. It was a summer morning of 1966 when an unknown person met me in my office at Calcutta. He identified himself as a Principal of a district school. He had just seen the new museo-bus at another school and decided to come all the way to Calcutta to demand that the bus must hold its exhibition at his school. To accommodate his request we had to change the entire schedule. The exhibition was over at his school in three days and the Principal visited me once again. He appreciated the mobile exhibition but pointed out that its effect would be short lived. The bus will visit his school probably after a year and nobody will remember its impact on the young minds. He wanted to sustain the impact and, for that matter, offered a 1,000 sq ft hall in his school, rent free, for setting up something that was later called a *District Science Center.* That was the beginning of the empire of what is now known as the National Council of Science Museums, with 26 science centers all over the country.

In the winter of 1965–66 a teacher of a local girls school in Calcutta came with a practical problem. She was facing difficulties in explaining solar and lunar eclipses on the blackboard and wanted to know if we could demonstrate these phenomena with the help of a three-dimensional model. While our museum was still working on this problem, another teacher approached us for help in demonstrating the difference between alternating and direct currents through a 3D model showing the functions of slip rings and commutator. These demands led the museum to start a new activity, the *School Demonstration Lecture*, with inexpensive teaching aids on a series of curricular subjects.

For school demonstration lectures, low-cost teaching aids were developed with scrap materials, with the help of hand tools and without requiring high technical skills. The expectation was that, after observing the demonstration lectures, science teachers in schools would be inspired to develop such low-cost teaching aids with the help of senior students and use them in classroom teaching. In the winter of 1967–68, when I was expressing this hope before a small group of science teachers in a district town, they all laughed at me. They said they did not even know how to properly use a screwdriver, because the formal education that they had taken did not have any practical component. How could the science museum expect that the teachers would virtually run a small workshop?

This led us to develop our *Teachers Training Program* and *Creative Abilities Center* in schools. It is something like an equilateral triangle with three activities on the vertices. First, science teachers would be trained in the use of hand tools for development of prototype teaching aids, then they would develop such teaching aids with the help of senior students in *Creative Abilities Centers* set up in schools and, finally, use the aids in classroom lectures for enhancing the quality of science teaching. It was a real requirement in rural schools, most of which did not have a science laboratory. This was the beginning of the *Build Up Your Own Laboratory Program* in India.[11]

That is the kind of story that can be told for most of our outreach programs developed on demand from our users. In order to make such activities really minds-on, the design considerations of activities need to be based on certain objectives which can be classified according to Guilford's five mental operations as shown below:[12]

(a) Cognitive—involving multiple-choice tests to recognize, realize, be aware of, become acquainted or familiar with or be conscious of certain facts, concepts or processes as reflected in the subjects of displays or activities.
(b) Memory—helping the audience to remember, acquire distinct or thorough knowledge of, or learn thoroughly the subject of displays or activities.
(c) Divergent—encouraging independent thinking through trying out different hypotheses, spirit of inquiry, creative and constructive approach, free response.
(d) Convergent—studying the behavioral norms, adoption of the right attitude in reaching the right solution through the process of elimination based on a logical sequence.
(e) Evaluative—process of critical thinking, assessing, selecting, comparing, judging and decision making.

One or a combination of the above mental operations can form the conceptual basis of the minds-on activities. The activities can be conducted through the ask-and-guess method, in which the audience would formulate hypotheses on the basic questions and then their responses are evaluated on a number of Guilford factors such as:[13]

(i) Sensitivity—response to problems and readiness to handle them.
(ii) Ideational fluency—proliferation of ideas, questions, guesses, improvements and solutions.
(iii) Flexibility—variety of questions, hypotheses and attempts for solution and improvement.
(iv) Originality—communication of new ideas, unusual consequences and causes, uncommon improvements.

CONCLUSION

Writers can use a thousand words to convey their message to a large number of readers. Artists communicate with their brushes. Films and video programs are used for mass communication. All these modes of communication are essentially passive and move in a one-way direction from the creator to the audience, with little scope for instant interaction. Science museum curators communicate through exhibits and activities—that is their language, their tool, their medium. And it is a very communicative language, a sharp tool and an effective medium. It provides for a close interaction between the audience and the object, a psychological involvement through participation, an all-pervasive action all over the floor of the museum.

It is important to remember, in this connection, that systems of thought nowadays comprise not only speculative or abstract concepts, but also those of a more empirical and matter-of-fact nature. We must not forget that cells are being manipulated to create life without a male partner, images of distant galaxies are analyzed to discover other planetary systems where one day we may be able to locate intelligent life forms and the whole of King Solomon's mine of information and knowledge is at our finger tips through the Internet. The science museum is the only place where human attainments can be seen and perceived under a single roof. Still, it is not just the contents of exhibits but the form of presentation that will develop a logical thought process in the individual mind and create a *scientific temper* in the community. Science museums will develop themselves into a new entity not for the men of science but for others. Keats lamented over the unweaving of the rainbow by Newton through a prism, but a modern scholar writes in his *Unweaving the Rainbow:* "the case of DNA finger printing suggests that lawyers would be better lawyers, judges better judges, parliamentarians better parliamentarians and citizens better citizens if they knew more science and, more to the point, if they reasoned more like scientists."[14]

NOTES

1. *Srimat Bhagabat Gita*, chapter 2.
2. "Foundations of Consciousness: How Ideas Bend the World into Shape," http://www.mind-field.com/
3. Samkhya, Patanjala, Nyaya, and Vaisesika are four schools of philosophy that flourished in India during the period between the 6th century and 2nd century B.C.
4. B. N. Seal, "The Physico-chemical Theories of Ancient Hindus," *History of Chemistry in Ancient and Medieval India*, ed. P. Ray (Calcutta: Indian Chemical Society, 1956).
5. *op. cit.* 2.
6. William Calvin, "The Cerebral Code" (Bradford Books 1998), p. 61, quoting Karl Popper.
7. Saroj Ghose, "Museums of Tomorrow: A New Look at the Concepts, Contents and Techniques," *New Visions for Museums of the 21st Century, A Forum of Museum Directors*, National Museum of History, Taipei, November 1998.
8. Saroj Ghose, "Interactive Participation in Science Museums," *Physics Education*, vol. 6, no. 3 (1989): 236–242.
9. Erich Harth, "The Creative Loop: How the Brain Makes a Mind.", www.amazon.com
10. *op. cit.* 2.
11. Saroj Ghose, "On Demand," *40th Anniversary Souvenir of Birla Industrial & Technological Museum*, Calcutta, 1999.
12. J. P. Guilford, "Creativity," *American Psychologist* 5 (1950): 444–454.
13. E. Paul Torrance, "Guiding Creative Talent" (Prentice Hall of India, New Delhi, 1969).
14. Richard Dawkins, "Unweaving the Rainbow: Science, Delusion and the Appetite for Wonder" (Houghton Mifflin Co., 1998, p. 113.

In Favor of Scientific Knowledge

THE NEW MUSEUMS

Jorge Wagensberg

1. THE CONCEPT MUSEUM (two preliminary principles)

Every form of transmission of knowledge has its own particular carrier element and others that are accessories. Thus, classes, lectures, debates and seminars are based on the *spoken word;* in cinema, television and other audiovisual media, the basic carrier of knowledge is the *image;* for books, magazines and other publications the essential element is the *written word,* the text; and play, which is used so much in the animal kingdom for learning, makes use of the device of *simulation*. And what is the case with museums? Museums, exhibitions and demonstrations focus above all on the *real object* (or *real event*). Each of these forms of transmission may borrow, as supplementary aids, other carriers of knowledge from the other forms. Thus, lecturers back up their words with transparencies (images), and museums explain their real objects with texts that they often complement with drawings or photographs.

One of the most ingenuous bad habits of today's museography consists, precisely, of forgetting this unrenounceable priority of the real. Exhibitions are reduced to a book with its pages blown up and arranged vertically, which the long-suffering visitor reads while walking along a passageway; museum rooms, which are packed with videos and computers, bear more resemblance to an electrical appliances store; and demonstrations are carried out on models of all kinds. To adopt the working hypothesis that a museum is based on real happenings and events would seem to be prior to the adoption of any other hypothesis. To do otherwise is to be placed at a disadvantage and to be left with exhibitions that are taken in by standing and reading, rooms filled with computer or video screens, and demonstrations of special effects. Let us accept the hypothesis, then. Reality will, at the very least, be a promise that will help us leave the house.

From this first principle arises a second one, with its own vigor. A glance at recent scientific museology will suffice to reveal it. In effect, it is undoubtedly the science centers and museums that have most changed their content, their methods, their role in society, and their attitude to the public:

Do not touch has now become *Please touch.* The concept of the display case has changed into the idea of the experiment; the idea of the academic label has turned into the text with literary quality; using the sense of sight as the only method of perceiving has given way to using all the senses (or almost all); the emphasis has shifted from the preparation of answers to the preparation of questions; and love of the past alone has given way to a desire to use the museum as a *tool for change*.

All this may be summed up in a second principle concerning the priorities of museology. Although nothing of what follows is prohibited, the first priority of a modern science museum is not so much to conserve a heritage, to inform, train or teach—the first priority is to provide stimuli. Why stimuli? What kind of stimuli? How are they to be presented in a museum? By combining these first two principles, we may venture an initial conclusion:

A modern museum should further the provision of stimuli based on objects and phenomena in reality.

However, this begs a question. Why a museum? Don't we have enough stimuli presented in formal education, in the established media, and elsewhere? The root of this question, the need for the museum and its stimuli, stands as a major contradiction to our present-day society. As follows.

2. THE GREAT CONTRADICTION (or the need for stimuli)

After thousands of millions of years of biological evolution, *natural selection* has consecrated a few extremely important basic functions such as *feeding*, *respiration*, and *reproduction*. How has evolution managed things so that the creatures of this world do not forget to eat and drink, to breathe, or to leave copies in time to continue the species? This question is in no way a trivial one. In fact, as everyone knows, matter in general—and living matter is no exception—is essentially lazy and tends always towards situations of minimum energy output. One might say, in principle, when it is a choice between doing and not doing, it is better not to do. Why leave the safe shelter for the hazardous outside and use up energy searching for food or a possible mate? If it were not for the pangs of hunger, we would not eat. This is what happens as evolution unfolds: certain powerful *stimuli* are *selected* in order to guarantee the basic functions—hunger and thirst for nourishment; a powerful, peremptory need for respiration; an efficient sexual stimulus for reproduction. To put it another way, creatures born with no appetite in one of these respects disappeared long ago. These stimuli, which are both urgent and not lacking in some promise of pleasure, would seem to be a necessary condition for the continuance of life. Now it has befallen us, the human race, to experience a real transhistorical ephemerides. The point is, we have invented (or maybe only consecrated) *knowledge* as a useful function for survival. Thanks to knowledge we have conquered the planet at a dizzying speed. After all, we gained access to it barely a hundred thousand years ago. What is the problem? Well, the problem is that natural selection has not yet had time to work in favor of knowledge in order to consolidate it with a stimulus at once urgent and pleasurable. This is at the heart of the biggest contradiction of civilization: we have become addicted to knowledge and selection has not yet given us a stimulus to consecrate it. So we cannot sit and wait for biological evolution to resolve our contradictions. There is only one solution: to give ourselves certain stimuli to further a thirst for knowledge and the creation of scientific opinion.

Perhaps I have exaggerated somewhat. A brief review of the history of the human race will suffice to show that, oddly, the most remote and separate civilizations have beliefs about the beyond, gods, rituals and all kinds of objects and artistic representations, and weapons, tools and protoscientific practices. According to some authors, around one hundred thousand religions have emerged in the last sixty thousand years. Let us make finer distinctions. Indeed, there exists a certain strong innate stimulus to knowledge of the revealed type. Otherwise, why is it so difficult to find any traces of human groups that do not display the worship of some divinity? There also exists some stimulus, although not so strong, towards art. This makes it possible to put the question in just the right terms: What does not exist is, above all, a stimulus that would further the spontaneous practice of the scientific method.

A few individuals, at great speed, produce huge amounts of scientific knowledge which is paid for, suffered, and enjoyed by the whole of humanity. This contradiction becomes more gigantic by the minute for two reasons. The first is obvious: how can the citizens of a democratic society that is marked by science influence their future if their level of scientific training is not even on a par with the level prevalent in the Middle Ages? They need stimuli. The second reason is not so obvious. When faced with any complexity, modern citizens usually turn to methods of an artistic or divine nature, for which they indeed have certain natural stimuli. The inventory is clearly incomplete. Let us look, for example, at something as common and at the same time as complex as understanding our fellow beings (from our own family, another family, another neighborhood, another city, another region, another country, another continent, another culture, another religion, another race). The history of human iniquity is the history of intolerance. And what is needed is not only tolerance but genuine acceptance of our fellow beings. The scientific method, if not scientists, can perhaps help towards this. This method, which, let us insist, not even scientists apply outside their work (often not even within it), is the only one that changes the truth when truth is no longer compatible with observation of the world, the only one for which everything is revisable.

Scientific stimuli are then doubly necessary: (a) for science itself and (b) for everything that, while not scientific, may benefit from the scientific method. With this we are even closer to the crux of the matter. A museum that accepts the provision of stimuli for citizens as its first priority focuses its work on creating a difference between before and after a person visits the museum. It is highly possible that one visit may not serve for much more. But that difference should serve for everything else: to promote reading, classes, training, information, the valuing and protection of all kinds of assets, and so many other experiences in favor of scientific knowledge. What is scientific knowledge?

3. CREATING SCIENTIFIC KNOWLEDGE (or the need for the method)

There are methods for creating knowledge and methods for transmitting knowledge. Probably, in the absence of more information, the best method for transmitting a certain piece of knowledge is the same as the one used to create it. It is even highly possible that the idea of teaching, understood as a set of techniques for transmitting knowledge of any type, simply makes no sense. This consideration may be applied to the idea of the science museum. In fact, the museum is for transmitting science and its content, science, has a specific method with its own particular tradition.

The method

Knowledge is a representation (necessarily finite) of a piece of reality (supposedly infinite). Science is knowledge elaborated with the scientific method. And the scientific method is any method that respects three principles: *objectivity*, *intelligibility* and *dialectic*. One is objective when, faced with various ways of observing an object, one opts for that which least affects the observation. One is intelligible when the representation, in some sense, is more compact than that which is represented. And one is dialectical when the knowledge risks being demolished by experience. Knowledge is scientific when it has the will to be so, that is, when it achieves the maximum objectivity, intelligibility and dialectic, however small these maximums are. According to this, a billiard carom mechanic can be just as scientific as a quantum physicist. According to this, there is no reason for a psychologist to be any less scientific than a physicist (that they may give up explicitly is another matter). In the same way, there is nothing against politics, a form of knowledge devoted to organizing coexistence, being built up with scientific method (that it may have made little effort is another matter). There are as many forms of knowledge as methods for making representations of reality. However, it is very possible that all of them are combinations of only three. In fact, in addition to scientific knowledge, there may be said to be two other forms: artistic knowledge and revealed knowledge.

Artistic knowledge does not necessarily require objectivity or intelligibility. It is based simply on a single amazing principle: certain infinite complexities, not necessarily intelligible, are transmittable through a finite representation (a musical score, a painting, a grimace). Revealed knowledge, for its part, is based on two efficient principles: (a) a being that possesses the knowledge of all reality exists, and (b) this being sees fit to reveal to us (sometimes) (part of) its knowledge. (Religion, that rare inspiration, and superstition are some of the means by which revealed knowledge is transmitted.)

Scientific knowledge therefore requires a method. However, this begs a question. Why do scientists do science? What, if any, is the need for it? What is their stimulus? This point, as we shall see, will be essential when we return to the subject of the transmittability of knowledge.

The stimulus

What is it that moves a scientist to do research? I do not believe that ethical reasons need be sought to answer this question. The scientist seeks neither the good nor the evil of humanity. It is rather more simple. The scientist, like any other citizen, needs to produce knowledge about the world in order to share, to the extent that this is possible, his or her cosmic loneliness. Science differs from other forms of knowledge only by virtue of the method used to produce such knowledge: the scientific method. Is there anything in this method that may stimulate the work of the researcher? The *experiment* has the leading role in the scientific method. Experimenting is an attempt at dialogue with nature. Not all the questions are good, and it is not even clear that questions should be asked at all. So there is not always an answer or, at any rate, there is not always an answer that will serve to produce knowledge. But when there is, when nature suddenly responds with something intelligible, then this is the scientist's moment of truth. It is the moment when the human being–nature communication is consummated, it is the emotion of the scientist, comparable, moreover, to the moment of emotion in art, that is to say, when the creator of a work communicates with an observer through a work. Stimulus and emotion are the key words for the zero point of all knowledge, the starting point for its creation or transmission. They are stimuli and emotions that result from experimenting with the world.

4. TRANSMITTING SCIENTIFIC KNOWLEDGE (in a science museum)

Let us recapitulate: (a) knowledge is an achievement of biological evolution, but can be transmitted by a route other than genetic inheritance, (b) stimuli are the starting point for the creation of knowledge, and (c) knowledge is elaborated by a method. In particular, scientific knowledge is elaborated with the scientific method.

How to choose stimuli? Amusement? Play? The benefit of the result? A prize of some kind? Fear of punishment or of some negative consequence? In fact, there are many kinds of stimuli, but, in the absence of other information, why not try the stimuli that make the scientist do science? In other words, the stimuli of the scientist have already served some purpose: to do science. They probably also serve to transmit it. Here we have a proposal for a working hypothesis:

I. The stimuli that further the creation of knowledge are the same that further its transmission.

In order to transmit scientific stimuli, it is preferable to feel them first. And the same may be said, I feel, of art or mysticism of any kind.

Second question. We have now succeeded in transmitting the stimuli and the recipient is thirsty for science. What method do we use to transmit the knowledge? How do we make it intelligible? Here is the second basic hypothesis:

II. The method that furthers the transmission of knowledge is the same as that which furthered its creation.

In a few words, seeking for stimuli in science itself and showing them objectively, intelligibly and dialectically is, at the very least, something to be tried. We now have a museological method to use to work out anything that a modern science museum may offer. Simply put, that would be any method that respects the two principles set forth. Taken together, these two working hypotheses are equivalent to a whole way of transmitting knowledge. It could be expressed more or less as follows:

The fundamental idea for the transmission of knowledge consists of the tendency to put the recipient of the transmission literally in the skin of the person who elaborated it.

In the jargon of science museologists and, by extension, of all museologists, a term has emerged to consecrate the method: *interactivity*. It is the way the citizen experiences the emotions of the scientist. The proximity between museologists and professional scientists (active researchers) is therefore highly recommendable. It should be observed, for the moment, that bearing in mind the hypotheses put forward, it is very difficult to be mistaken and confuse it with the nature of the spectacle that a science museum should offer. The purpose now is to seek within the emotions of a scientist and try to be a little more precise.

5. THE METHOD OF EMOTIONAL INTERACTIVITY (exhibitions)

After thirty years of history involving science museums, I believe we have come a sufficient distance to be able to state that there are at least three different aspects to the involvement of the visitor in a museum. With a good combination of these three ingredients, to be determined according to the nature of the real object or happening to be transmitted, what we might well call the total interactivity method is defined. The three ingredients are:

1. *Manual interactivity* (*Hands on* in museological jargon). The first kind of scientific emotion is based on the experiment. The visitor is an active element in the experiment, uses his or her hands to provoke nature and watches with emotion the way in which nature responds. This type of interactivity, more specifically called Hands on, has in only a few years become the obsession of all museographical scriptwriters. The idea of Hands on has now gone through many phases: its beginning or the revolution; its consecration or splendor; its consolidation but also its vitiated application or degeneration. The lowest point of manual interactivity is in the semi-deceit of simply starting (a pushbutton that starts off a show that is otherwise totally passive). And its high point is when the citizen can use the response of his or her action to initiate a new action, when the visitor really asks questions and when it is nature, with no intermediaries, that responds. The vitiated aspect of manual interactivity is when all that is sought is for the visitor simply to do something, anything—whatever it might be. For example, the visitor might push a button to light up an object that, by default, would be in darkness. The point is that manual interactivity may not amount to much without a dose of mental interactivity.
2. *Mental interactivity* (here let's call it *Minds on*). The visitor's mind experiences a clear change during the period between before and after the visit to the museum. The mind comes up against a challenge that gives it work to do, for example, to feel a desire to ask one, or several, new questions. To come out with more questions (even with more doubts) than when one went in is, precisely, a good gauge of an exhibition's value. To have something to solve, to come upon a new analogy, to spot a paradox or a contradiction, to glimpse a new idea, to be hit by a new suspicion, to register a new piece of data, to plan a new experiment—all this triggers activity between the mind and reality. But there can be no good mental interactivity without a dose of emotional interactivity.
3. *Emotional interactivity* (now let's call it *Heart on*). In an exhibition, the time spans available to focus the visitor's attention are short. For the visitor's mood or humor to receive some kind of emotional discharge, the most genuinely cultural aspect has to be approached. The object or happening on exhibit may give hints of an aesthetic, ethical, moral, or historical nature or simply of everyday life, which makes a connection with a sensitive facet of the visitor. This is when it becomes legitimate to use art to communicate science. It is, at bottom, what makes it possible for every museum to be different.

Every good museographic stimulus is based on a good combination of these three ingredients. Strictly speaking, the three types of interactivity are emotional, which is why I would call the total interactivity described emotional (in the widest sense) interactivity. Let us look at a couple of examples taken from a room at the Museu de la Ciència de la Fundació "la Caixa" in Barcelona. These are (a) *Invisible stillness or the art of disappearing* and (b) *Very brief history of three rocks born different.*

Invisible stillness or the art of disappearing. The visitor approaches a large terrarium, perhaps having

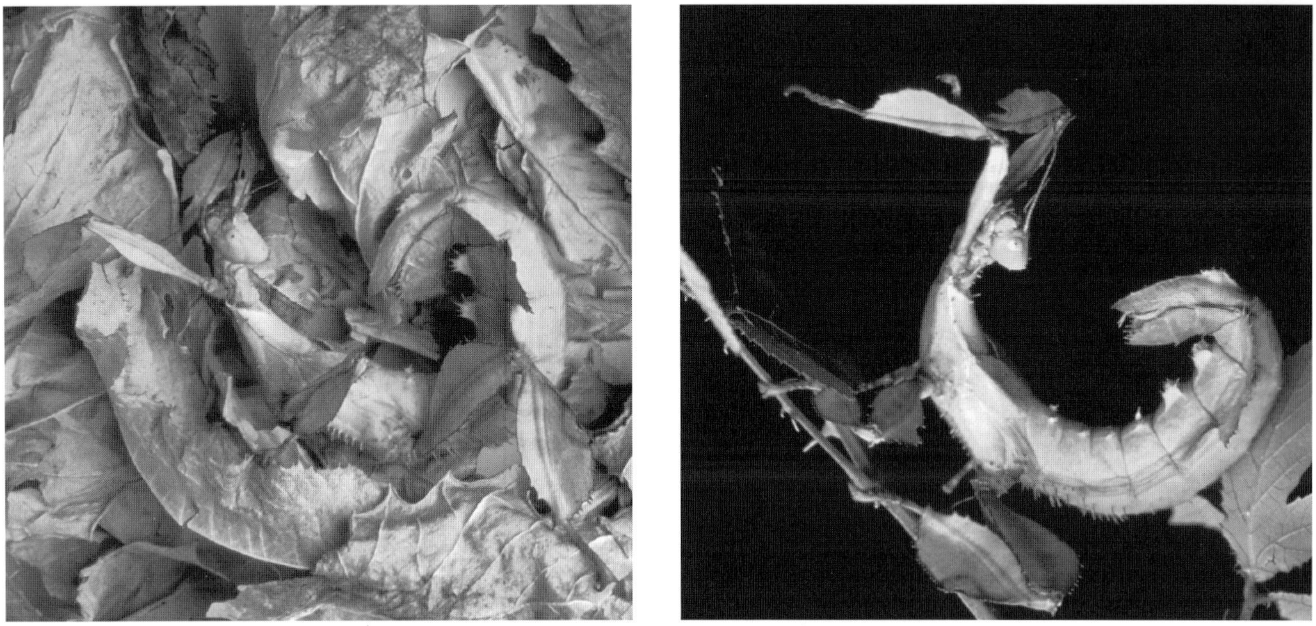

FIGURE 1. A stick insect in two different landscapes: a real object stimulating the "minds on."

been attracted by the title. The first thing he or she sees is nothing. Because in the interior, decorated with natural materials (leaf litter, earth, roots, a stream, etc.), there live no fewer than two or three dozen stick insects. The visitor's first emotion is, in every case, that there appears to be no emotion. The eyes run ironically and hastily over the small space. Another module closed for repairs! The visitor looks but sees nothing, until he or she comes across the notice *Here there are 30 large insects*. How can 30 insects not be seen in such a small space? The visitor's incredulity grows, but her concentration also grows, and suddenly . . . aha! She sees one clearly, two, three—the terrarium is full of them! Before that moment her eyes had been looking at them, but her brain had not seen them. This is emotional interaction. Most visitors are literally grabbed for the perception of all that follows. Next to the terrarium is a window through which one perceives a cloud of haphazardly arranged dots on a plane. It is impossible to recognize the slightest meaning in the arrangement. However, if the visitor activates a control, some of the dots move and the picture of an animal appears in sharp focus. The manual interactivity is genuine: when the action is performed, the animal becomes visible although it was already there, but, when the action ceases, the animal, while still there, disappears right under the visitor's nose. The phenomenon stimulates the imagination. It is the generic concept of the extra. Many supposed victims adopt the strategy of rigorous immobility even though the predator is already breathing down their neck. Furthermore, this links up with everyday life. Why do we wave our hand when we want to catch the attention of a waiter who is an expert at turning a blind eye to everything out of his direct line of vision? Analogy! Many other previous experiences are reinterpreted, reread. This is *mental interactivity.*

Very brief history of three rocks born different. Before entering the museum, the visitor comes across three magnificent hexagonal basalt columns fashioned like those of a sculptural-totemic monument. If there is a queue at the ticket office, as is usually the case, many people wander over to have a look. An enigmatic plaque says *Liquid rocks: Basalt columns from Castellfullit de la Roca. See experiment in the Living Planet Room*. It is highly likely that the *emotional interaction* is triggered in the artistic sensibility of many. And not only artistic. There is also some likelihood of touching national pride, a likelihood certainly proportional to the distance between the area frequented by the visitor and the lovely town, near Barcelona, that the rocks hail from. So we start off with a touch of *emotional interactivity*. Once

FIGURE 2. **Captain's room "before and after" the wreck in the entrance of the exhibit "Huracan 1724." Museu de la Ciència de la Fundació "la Caixa."**

in the room mentioned, the visitor comes upon a threefold experiment presided over by three different authentic rocks: (1) a piece of hexagonal basalt column; (2) a sedimentary rock cut cleanly to show the progression, from smaller to larger, of the size of the grain; and (3) a volcanic rock, similarly cut, in this case, to show the arrangement of the pores, from smaller to larger. Each of these rocks presides over an experiment that the visitor can provoke, that is, through *manual interactivity:* (1) a race of different-sized particles in a viscous medium to show how they order themselves, like the grains of sedimentary rocks! (2) a race of air bubbles in a liquid to show that they arrange themselves like the pores of a volcanic pump! and (3) a spectacular Rayleigh-Bénard convection to show that the particles are arranged in a honeycomb structure like the basalt columns, on one of which, without first realizing it, the visitor has propped his elbows to have a better look! And the *mental interactivity* is unstoppable. The three examples are three explanations of three objects, but explained by nature itself at the visitor's request. An object is an objective, intelligible and dialectic result of a process. The analogical mechanisms also start to work. How many scientific disciplines are devoted to reconstructing history? Geology is one of them. But there are many other sciences that focus on the study of remains and traces: paleontology, paleoanthropology, archaeology, cosmology, history itself, police detective work, and so on.

The need to create stimuli to further scientific knowledge, as aim, and emotional interactivity, as method, mean that now we can think about designing and constructing a good science museum. Now we have permanent and temporary exhibitions and, therefore, according to the result of our work, we shall also have a degree of prestige, of atmosphere, and of credibility. If all this reaches a certain level, then it is time to think of the science museum as a civic institution. For example, can the museum fulfill any other function? What is its role in a modern democratic society? The museum serves, at the very least, to help the citizen create scientific opinion. An event that took place a few years ago at the Museu de la Ciència convinced us just how relevant this is.

6. THE CREATION OF SCIENTIFIC OPINION (at a science museum)

Barcelona, 7 P.M., Tuesday, 12 April 1992. Main lecture hall. Today, as part of the *Evenings at the Museum* series, three important scientists from the Spanish Research Council have come to debate with the audience on the subject: *The Mediterranean is not yet dead.* There is an environmental chemist and director of a research center, an oceanographer, also the head of a center for advanced studies, and a research biologist from the Institute of Sea Sciences. In a word, official science. The lecture hall, which seats 200, is packed, and so is the supplementary room for 150, where the session may be followed on closed-circuit television. The anticipation is enormous. My job is to introduce everyone and moderate the debate. One by one the scientists produce their facts and set forth their arguments with the aid of a wealth of up-to-the-minute data, neat graphs and attractive photographs. A feeling of optimism begins to pervade the hall. For example, a measurement of the rate of renewal of the Mediterranean waters is defined. From the time a molecule of water comes in through the Strait of Gibraltar from the Atlantic and leaves again, one hundred years elapse. This is the fact. The interpretation that the scientists give is that the Mediterranean is very large, in many respects it has the characteristics of an ocean. Another example: there are tons of crude oil lost in the sea for a variety of reasons. This is the fact. The official interpretation is that this is a very small matter. There are appropriate bacteria that will sort it out with no problem. Another interpretation is offered: the constructions for oil prospecting do not represent any extra tension for deep-sea flora and fauna. In fact, quite the opposite: they represent artificial protection that will further the blooming of new underwater havens. Not only is the Mediterranean not yet dead but it would also seem to be no easy matter to do away with it. Everyone seems to be in agreement, the scientists and the audience. The evening is unlikely to end in debate. Indeed, they look at each other in pleasant surprise. But the moderator, despite his naturally naive and cheerful character, scrutinizes the audience a little crossly. There is something fishy here. Clearly the audience was convinced beforehand. They are from the social milieu of the speakers: co-workers, students, relatives . . . except for the back row! Something is going on in the back row. The moderator recognizes one of them: no other than the president of Greenpeace in Spain. This is much better. The moderator's memory suddenly seems to have been jogged, he glances at the program in front of him, and remembers! For the following day, 13 April, purely by chance, the renowned NGO had hired this very hall for a very similar function. The title was very similar to this one: *The Mediterranean is dying.* The coincidence was notable because, although it was a private session, I had also been asked to act as moderator. Now it is clear: the NGO people have heard about the session that was to precede their own and have turned up to check it out and present battle. That's why they are here. The moderator rubs his hands. When this happens, there is loud clapping and the moderator throws the debate open to the floor. Contributions from the floor are on the same wavelength as the talks—here a detail, there a flattering remark, a prediction, a confirmation. The moderator looks toward the back row with growing impatience. Aren't they going to say anything? Finally he addresses them with eyebrows raised and hands spread as if to say: go ahead, you obviously don't agree with what has been said, now I'm giving you the chance! But the eco-activists look blank and shrug their shoulders, turning down the invitation, as if to say: there is nothing to be done with these people, it's the same old line, it's useless even to try! Before taking my leave of the audience, I announce the debate for the following day and invite the audience to come and see what the other side of the coin might be. I identify the people in the back row and everyone turns around to have a look. After thanking them, I tell the scientists what has happened (the scientists are surprised at my surprise: what did you expect?) and I summon them for the following day. I even offer to buy them a meal (at the restaurant of their choice) if they will agree to come back, as audience, and kick off—they themselves!—the debate after the ecologists talk. They promise to do so amidst laughter and predictions for the next day.

Barcelona, 7 P.M., Tuesday, 13 April. Main lecture hall. Today, as part of the *Evenings at the Museum* series, the same moderator sits down to debate the same topic that was presented on the previous evening. Only the speakers and the audience have changed! Not a single member of the public from the previous day has come back. The two rooms are also packed, but this time with onlookers, sympathizers and all-out supporters of Greenpeace. After showing transparencies of dangerous acrobatics performed by members of this NGO, which the audience applauds frantically, the papers on the Mediterranean begin. And they start the same as the previous day's. The same facts. The residence period of a molecule in the Mediterranean is one hundred years. They agree about that. But today this means that not only is the Mediterranean not an ocean, it is a pond! The same fact but a different interpretation. There is also agreement about the volume of crude spillage into the sea, but disagreement about what this figure signifies. Bacteria digesting the oil? In the Persian Gulf maybe, but not on the Costa Brava or the Côte d'Azur. Havens beneath the platforms? You must be joking!

Needless to say, these two sessions suggested an idea: that there be a third session open to the two factions and to all audiences. This took place a few months later. And there was a debate. And fine distinctions were made, because, as is well known, the essence of any debate is a question of fine distinctions. But the exceptional significance of this incident has been—and we intend it to be—much more far-reaching.

The anecdote marked a watershed in the history of the museum, a before and after. It illustrates a failing of our society and is the foundation stone of an entire proposal. Analysis is not difficult:

1. We have a problem. Scientific questions affect not only the scientific community. In a democratic society, the possibility of taking part in decision making should be open to everyone. And, increasingly, any decision requires prior scientific knowledge. Scientific knowledge is above all the realm of scientists, but scientific opinion should be equally rendered by any citizen. Where can science for citizens be done? The museum, as an institution at the same time scientific and civic, would seem to be a good place.
2. The same scientific data—objective, intelligible and dialectic—admit different, even contradictory interpretations. So it is not enough to transmit the results of science. Perhaps it would be a good idea to transmit, too, the method used to obtain them; perhaps it would be a good idea to work expressly and especially to further the creation of scientific opinion: the opinion of the scientist and the opinion of any other citizen. And all this is part of the vocation of a science museum.
3. The scientist, in particular, is a citizen like any other, with his or her (legitimate) ideology, areas of a priori sympathizers, with the right to make mistakes and the duty to generate scientific opinion. So, in an initial phase, it would be appropriate to spark discussion between scientists (which apparently does not come about

spontaneously often enough). And then also to spark discussion in an atmosphere with a greater mix of thinkers and different tendencies. The museum is the ideal place for both. Probably it is the only one. In effect, the universities and research institutes are atmospheres where anyone, except scientists, feels inhibited. Places attached to prior ideologies (however legitimate these may be) will arouse the suspicions of scientists. However, with prestige and credibility gained the hard way, the museum can attain acceptance by all as a good setting for discussion.

Conclusion: a modern science museum has, at the very least, two things to offer: (1) permanent and temporary exhibitions, in principle on reality in its entirety (from the quark to Shakespeare) and (2) within the framework of what has been achieved with the former, an intensive program of activities intended to directly foster scientific opinion: courses, lectures, seminars, conferences, debates, and so on. In particular, little by little the museum should become the natural setting for a series of events. For example, any scientist passing through the city for whatever reason (in general to visit a colleague or attend a scientific meeting) who is suspected of having something of interest to tell the community should end up talking at the museum. In a city like Barcelona, a museum may bring a dozen Nobel prize winners in touch with the public every year. The activity generated by a museum, picked up by the media, multiplies the scientific life of a city.

It is easy to conclude that every city above a certain critical mass should demand a science museum, just as it demands a chamber orchestra or a symphony orchestra. This raises one final question: what should the role of a science museum in modern society be?

7. THE ROLE OF A SCIENCE MUSEUM (in the society of the 21st century)

In terms of scientific knowledge, the citizenry, the community of citizens, falls into four major sectors:

1. The sector in which scientific knowledge is created. This includes the universities, the research organizations, and the laboratories. *This is the scientific community.*
2. The sector in which scientific knowledge is applied. This includes industry, services, agriculture and fisheries. *This is the production sector.*
3. The sector that manages the first two, that is, the creation and application of scientific knowledge. This includes the academic authorities and the political leaders. *This is the government.*
4. The sector that suffers and benefits from the achievements of the creation and application of scientific knowledge. This is all citizens taken together. *This is all of society.*

The place that a modern science museum should occupy is clear in this context: simply stated, it should be a common natural stage for the actors from these four sectors.

Now the time has come to dream. Let us imagine that this is so, that the science museums are like this one day, for example in Europe. Or in the world. What we shall have in reality is a network of centers dedicated to scientific knowledge for the citizen; all about science, its results, and its methods; all sharing debates on everything regarding those results and methods that may affect the citizen. All the museums will be different, because museums are made by men and women, and men and women are all different. Aesthetic and artistic sensibility will change, many cultural aspects will change, beliefs will change, but, if they are science museums (and what is more, all linked up to each other) by definition and method, they will all have many things in common. And having things in common is something that aids living, living together and tolerating one's fellow beings. To dream of a network of science museums for the coming century is to dream of a network of a new kind of cathedral, with a network of fresh air.

REFERENCES

Jorge Wagensberg, *Ideas sobre la complejidad del mundo*, Metatemas series, 9. Tusquets eds. Barcelona, 1994.

Jorge Wagensberg, "Public Understanding in a Science Center," *Public Understand. Sci.* (1), 31–35, 1992.

Virtual Objects

THREAT OR SALVATION?

Doron Swade

A central article of faith in museum life is the belief in the primacy of original artifacts. Traditional museums are part of an object-centered culture, and physical artifacts, their significance and care, dominate the curator's professional psyche. There is a new genre of object in the museological landscape—the virtual object. By this I mean an electronic representation of a thing whether real or imagined.[1] To what extent do virtual objects challenge the central article of faith that sanctifies the original artifact in museum culture?

Our relationship to physical objects remains mysterious and it is one that sociology, psychology, and philosophy have yet to exhaust. Objects are supposedly permanent. By this we mean that their rate of deterioration is slow compared to the windows of our own awareness. Objects outlast us. Perhaps possession confers a kind of immortality that defies time. Objects serve as messengers from the past—Grandad's chess set, Mother's wedding ring, Dad's pipe. Family heirlooms are imbued with associations and personal significance. We populate our lives with them. When we notice them, encounter, or contemplate them, they trigger internal dialogues with past owners and stimulate trains of thought as we visit and revisit memory, embroider and elaborate, explore and speculate. We bequeath objects to posterity as vessels of partially decoded meanings. Civil monuments and personal mementos serve as signifiers of ephemeral events and lives. At a personal level, possession and its satisfactions remain enigmatic and incompletely understood. The issue is a deep one, and despite our efforts at analysis, our relationship with things remains mysterious.

It is not clear whether institutional collecting is any better understood than collecting at a personal level. Yet Western society sanctions and endorses the act of collecting however well- or ill-understood the underlying reasons. One way society expresses the value it places on objects, their acquisition and preservation, is through the public funding of national museums. Museum-like institutions in industrialized Western societies are relatively well-endowed, though I understand that there are good reasons why museum directors who lobby governments for money might be reluctant to admit this to anyone but an intimate friend.

Before we can address the issue of whether or not virtual objects challenge the primacy of original artifacts, we need to articulate what it is about original artifacts that sanctifies them in museum culture. In the 1980s the United Kingdom underwent a cultural revolution. The comforts of the welfare state were subjected to the unforgiving scrutiny of the market. Under "entrepreneurial

Conservatism," accountability became the touchstone of efficiency.[2] The public sector, national museums included, was not immune. Curators were asked to justify themselves to newly appointed accountants and managers imported from the private sector. As a curator, I was asked by a new Director of Finance why the Science Museum needed original artifacts. Why he asked, could we not make do with replicas? To this day I ponder how an outwardly pleasant man could ask such a hostile question. It remains unclear to me whether the question was an ingenuous enquiry from an outsider to museum culture or whether the question was a disingenuous provocation—a flaunting of new power as a taunt to the newly disempowered curators. Let us take the question in good faith, however it was meant. How *does* a curator justify to a skeptic—someone who does not subscribe to the article of museological faith about the primacy of objects—the utility of original artifacts?

I offer to the Director of Finance the case of Napoleon's waistcoat button.[3] A curator is shown six objects. Because of financial constraints he/she has to choose one and only one of the six. The objects on offer are a button worn by Napoleon on his waistcoat at his final battle. Its provenance is impeccable and authenticated. The curator is also offered another button made by Napoleon's button maker but not worn by Napoleon at the final battle. The third and fourth objects are replicas: a contemporary replica of the button and a modern replica both physically indistinguishable at first sight from the original. The fifth object is a fake—a button claiming to be one from Napoleon's waistcoat but which was not. The final object is a hologram of the original button. The curator has but one choice. In choosing only one object I would hazard that the curator's instinct is to choose the original button worn by Napoleon at his final battle. The question is why?

Let us extend this scenario. An historian advances a new theory: that Napoleon was allergic to snuff; that he took snuff for the first time on the eve of his great battle; and his strategic judgement was impaired as a result. Now it does not make sense to examine anything but the original artifact for traces of snuff in an attempt to verify this thesis. It would be equally meaningless to examine the two replicas or the fake, since the issue of snuff content was unforeseen at the time of replication. Replicas only embody physical predicates seen to be meaningful at the time of replication. In the absence of a snuff thesis, any examination of anything other than the original artifact, however useful in other respects, would clearly be meaningless. Only the original artifact will suffice for this forensic purpose. I would therefore argue that the historical utility of original artifacts is that they are available for interrogation in the light of unforeseen enquiry. The original artifact embodies an inexhaustible aggregation of predicates the full meanings of which are perpetually incomplete. I would argue that acquiring the original artifact is the only course of action consistent with the open-endedness of human enquiry.

It could perhaps be argued that European history and an emphasis on great battles may not be an appropriate model for science museums. The allegory of Napoleon's waistcoat button is directly translatable into the history of science. Consider the case of Michael Faraday, the great electrician. This renowned Victorian experimentalist was subject to recurrent and mysterious illnesses. He lived and worked at the Royal Institution in Albemarle Street in London, and when he was ailing, his medical advisors counselled him to take the country air to recover. After a spell away, restored to good health, he would return to his London laboratory only to fall ill again. His symptoms are well documented. It has since been suggested that he was being poisoned by the materials (mercury, for example) he was working with. In the light of modern toxicology, it would be eminently sensible to examine his workbench, preserved in Albemarle Street, for traces of specific toxins. It would not make sense to examine a replica or a fake, since these can only reflect the state of knowledge at the time of manufacture. It is possible that the mystery of Faraday's illness could be illuminated by other means—contemporary records, contextual and circumstantial evidence. But what partly underpins the value of original artifacts and their mystique is the sense that understanding is never complete, that knowledge is never total, and that the original artifact embodies evidence the meanings of which may not be accessible by other means.[4]

The allegories of Napoleon's waistcoat button and Faraday's workbench argue the forensic utility to history of original artifacts. They also suggest that the mythological appeal of original artifacts might be underpinned by a sense of epistemic incompleteness that prompts us to endlessly visit and revisit episodes in history to reframe their meanings. In the history of technology, unchanging objects are the mute witnesses we consult each time in this unending cycle of inquiry. I would not wish to suggest that the forensic utility of objects in history is their only value. But framing the meaning of objects in this way is both a device to argue utility to a sceptic and also one that provides an heuristic tool to explore object surrogates.

With this in mind, we now have a basis for revisiting the question of virtual objects. We can now ask whether virtual objects, whatever their other merits and appeal, have consequences for the primacy of original artifacts as evidentiary source. Do they pass the test of historical utility? Can we interrogate them for answers in any meaningful way? I propose to explore answers to these questions using two examples. One is a computer simulation of a nineteenth-century calculating engine, a machine that the inventor and designer, Charles Babbage, never built. The other is a computer simulation of an electronic computer, the Ferranti Pegasus built in 1959. Each of these examples involve a physical object and an electronic representation.

Figure 1 depicts Charles Babbage's Difference Engine No. 2. The designs for this vast calculating engine that were completed in 1849 consist of some twenty large drawings that depict the mechanisms of the machine. The drawings are not without omissions, minor dimensioning inconsistencies, and other errors, but they do make clear how the 4,000 parts of the machine were intended to work.[5] The Engine remained unbuilt for over 140 years. In 1985 a Babbage scholar who had studied Babbage's designs suggested that Difference Engine No. 2 be constructed in time for the 200th anniversary, in

FIGURE 1. Difference Engine No. 2, 1989.

Courtesy Science Museum, London

December 1991, of Babbage's birth. In 1989 the first metal was cut. Up to that point Difference Engine No. 2 was a virtual object—a mental artifact of Babbage's inventive mind, depicted on paper. In November 1991 the calculating section of the Engine was complete and working at the Science Museum in London. The completed Engine was placed on public display and demonstrated live to groups of appreciative visitors. A virtual object had become real through a detailed program of construction carried out in public view. The Engine is not a reconstruction or a replica: it had never been built so it has no contemporary predecessor. It is a modern original of an old design.

The calculation mechanism is elegant and sophisticated. Though the elemental mechanism is inherently simple, it is repeated time and again in the full machine. The repetition of the basic mechanism, as well as the components required to drive and support the active elements, contribute to a visual impression of great complexity. To a visitor, the mathematical principle of the Engine (the method of finite differences) and the operation of the mechanism are not self-evident from a visual inspection of the physical machine. By way of an illustrative aid to viewers, a computer-generated simulation of the mechanism is provided alongside the Engine on public display. This animated electronic representation with a voice-over commentary offers the celebrated advantages of the electronic genre. On screen, freed from the physical constraints of real mechanics, parts of the mechanism are suspended in space without the confusing clutter of physical support and without drive mechanisms that distract from the essential logic of operation (a form of visual synechdoche). The mechanism is built up on screen part by part through a process of progressive disclosure so as to craft the explanation in an incremental way (Figure 2); the operation of small parts is magnified by zooming (Figure 3); critical functions are repeated to highlight different aspects; perspectives are altered to offer new viewing standpoints and detail of otherwise hidden wonders; and the rhythm of real-time operation is stretched, contracted, and frozen according to the explanatory needs of the spoken commentary.

The representations of the Engine's mechanisms on screen were not derived from the physical machine. Dimensions of parts taken from Babbage's original drawings were input into a computer program to define the individual components and the programmer instructed the program as to how the parts were to move and interact according to an understanding of the principles and intended operation of the mechanisms. So the simulation, derived

FIGURE 2. Computer simulation of Difference Engine No. 2.

Courtesy Science Museum, London

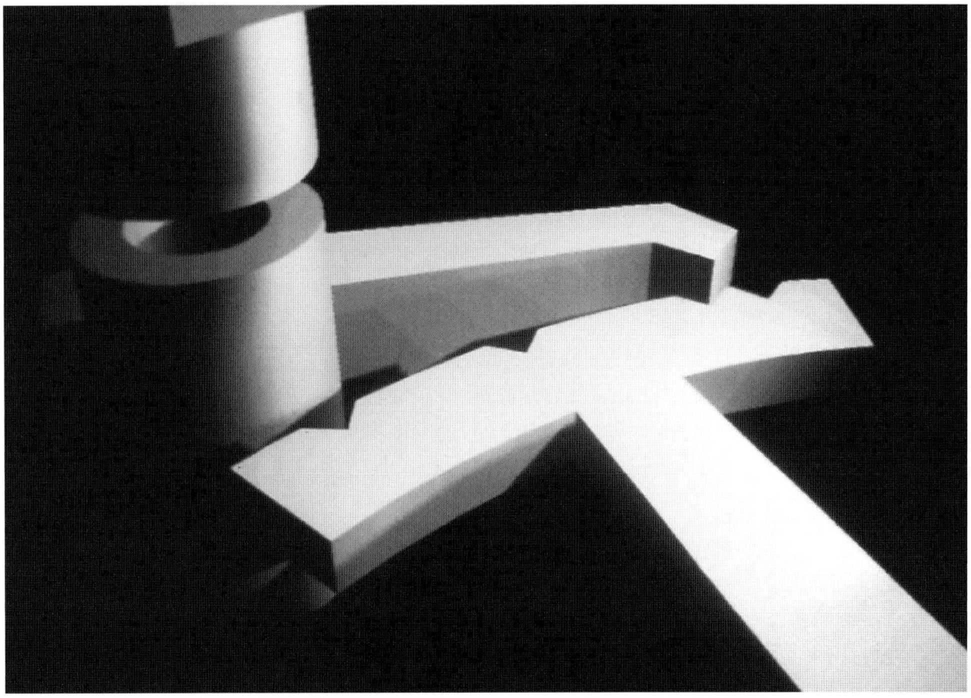

FIGURE 3. Computer simulation. Detail of carry lever.

Courtesy Science Museum, London

from the original drawings interpreted through understanding, exists quite independently of the physical machine.

The question we can now ask is whether virtual objects of this kind have consequences for the primacy of original artifacts? Clearly this simulation has educational and expository value undeliverable by the physical object. But does this virtual object undermine the primacy of the physical object as an evidentiary source? Does it pass the test of historical utility? Can we interrogate it for answers in any meaningful way? I would suggest that this simulation is no different in museological principle from that of the physical replica in that it only incorporates predicates bounded by the understanding of the maker—in this case, the programmer. Only predicates perceived to be significant at the time of "replication" are incorporated into the "object," and these are a reduced set of predicates at that (namely, the visual ones only). Such a simulation can be interrogated in much the same way as a physical replica, but only to reveal predicates seen to be significant at the time of replication. This suggests that whatever the educational and presentational value of such programs, whatever their considerable visitor appeal, and however promising their museological potential is in other ways, they do not have significant consequences for the primacy of original artifacts as evidentiary source in the forensic sense described earlier.

The freedom to represent the motion of parts independently of physical constraints derives from the separation of representation and thing. However, the grammar of electronic artifacts is unlike that of physical artifacts. Electronic representations are not bound by the physical laws of the objects or phenomena they depict: stones can be depicted falling upward; light can be portrayed bending around corners; objects can move without being acted on. In computer simulations, representation and referent are uncoupled. A human programmer instructed the computer to generate the images of the Engine. Just as craftsmen making a physical replica use their skill to minimize a physical difference between representation and thing and mediate between the two, so the programmer mediates between the object (or the specification of the object) and the computer-generated image.

In the case of the Babbage simulation, the uncoupling of representation and referent, which provides the basis for expository power, was found to carry its own unique hazards. In ordinary experience, it is impossible for two solid objects to occupy the same space at the same time. In the physical

Engine, the machine jams, fouls, or breaks. The behavior of electronic representations knows no such restriction. In the simulation of the Babbage engine, there are instances in which meshed gears slide through each other as would ghostly phantoms. There are levers that move without being activated by any causal connection with other levers, and there are instances of conditional behavior that is simply wrong. These are errors of misrepresentation. You can say, "program these constraints in." Precisely. To ensure conformity between representation and thing requires mediation by a human agent. Here it is the human instructing the machine. The value of original artifacts lies partly in their potential to instruct *us* by revealing the unknown in response to polite and sympathetic questioning.

There is one category of modern object for which electronic representation offers something that is both novel and museologically unique. For this category of object, the logical replica embodies an inexhaustible set of predicates and can be interrogated in the light of unforeseen enquiry in ways that physical replicas cannot. The category of object is computers, and the example used here is that of the Ferranti Pegasus.

The Pegasus computer shown in Figure 4 is a large vacuum-tube machine built in 1959. It was restored to working order by several of the pioneers who designed and maintained it, many of whom form part of the Computer Conservation Society founded in 1989.[6] Pegasus is the only working contemporary vacuum-tube computer I am aware of. It is a sad fact that however skilled the practitioners, and however conscientious they are in restoring and nurturing the machine, within about fifteen years Pegasus will be unmaintainable. When it comes to archaeological time scales, restoration to working order is no more than a short extension of its operational life. Figure 5 shows the console of the physical machine and Figure 6 a screen of the Pegasus simulation running on a modern personal computer. The simulation is what is called a bit-level simulation, that is, it is implemented at the logical equivalent of the molecular level of the machine. It replicates the behavior

FIGURE 4 Ferranti Pegasus computer, 1959.

Courtesy Science Museum, London

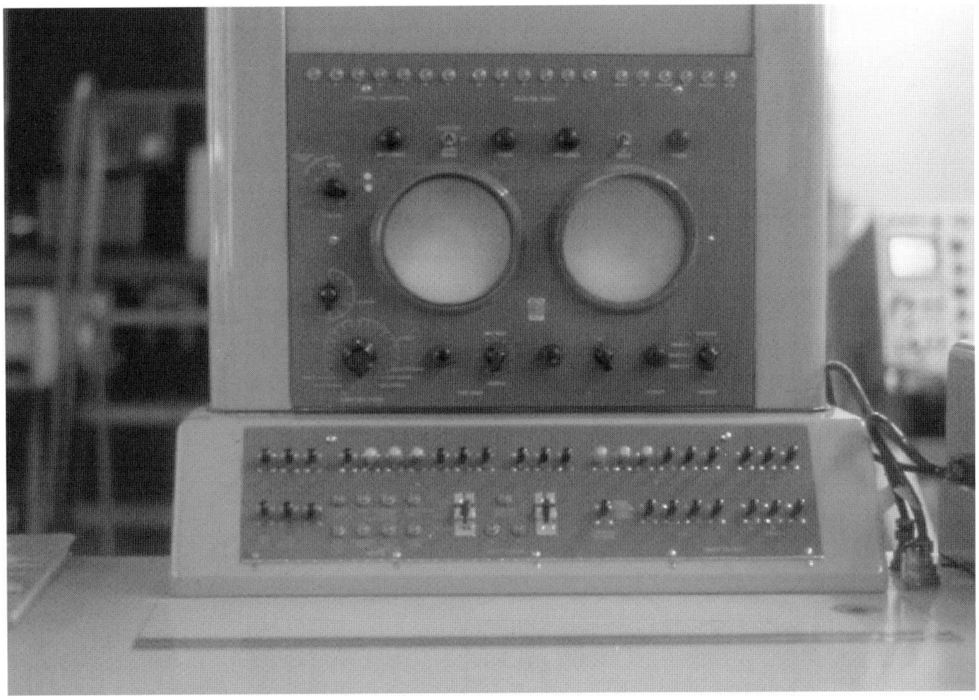

FIGURE 5. Pegasus: Engineer's console.
Courtesy Science Museum, London

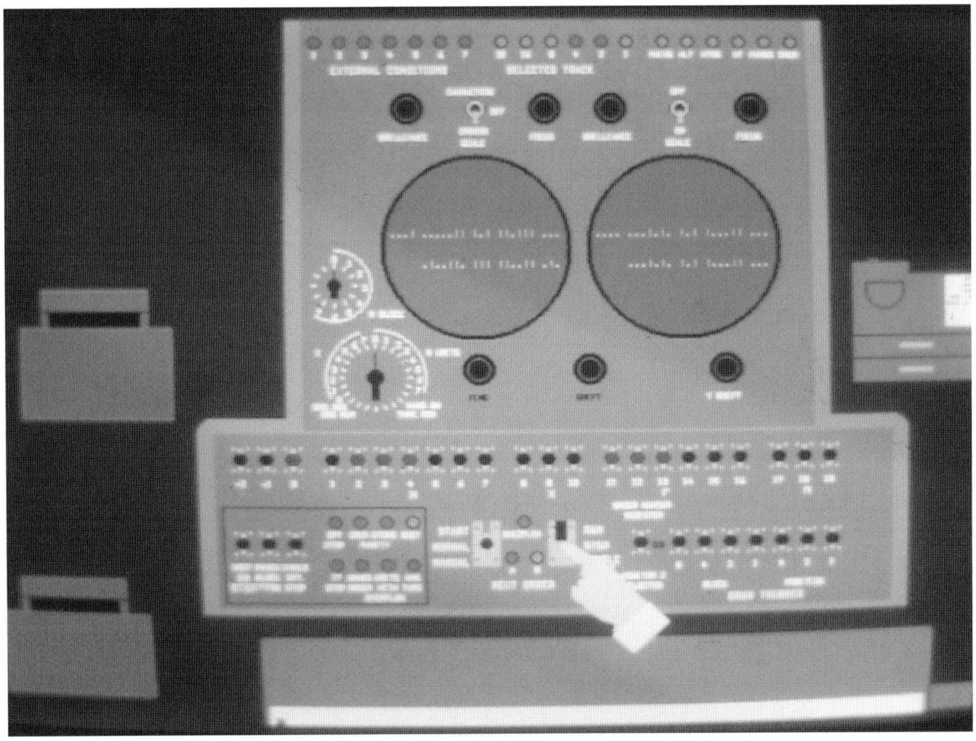

FIGURE 6. Simulation of Pegasus console.
Courtesy Science Museum, London

of the operating system exactly and animates the on-screen console in response. The issue is not how good is the visual facsimile of the texture, color, and form of the on-screen simulation, but the degree of completeness of logical replication, which is exact.

An intriguing feature of computers as museum objects is their artifactual identity. Because the world of museums is object-centered, we assume through habit and convention that what we refer to when we talk about Pegasus is the physical machine. In his famous "cold porridge" speech, Alan

Turing, the English mathematician and computer pioneer, argued that what defined a computer was not the medium of its physical implementation but the logical rules that define it. In a speech to the pupils of Sherborne, his old school, he likened the appearance of the human brain to the cold porridge he recalls being served at breakfast as a school boy ("grey and soft and wrinkled on top").[7] Despite the resemblance of brains to porridge, we do not hesitate to accept that brains think. Turing asserts that what is important is not the "wetware" of the brain, but the rules by which it operates and, by analogy, it is not the physical medium through which rules are realized that is significant, but the rules themselves. He argues that computers are universal machines that can be instructed to comply with any set of rules that can be described in symbols. One computer can therefore be fed the complete set of rules by which another computer operates, and one could argue that the identity of the machine, defined by and resident in its rules, has been transferred. It seems then that the identity of a computer is not exclusive to its physical hardware, which may be regarded as accidental to existence but is at least partly, if not wholly, owned by the logical rules that define its operation. In the case of computers, identity is abstract and apparently replicable without loss and without a reduction in the original predicate set.

If one computer is programmed at a sufficiently fundamental level to function in accordance with the rules of another, we have in fact an operational replica perfect in all logical respects to the original—even to the bugs and imperfections of the operating system. We can run programs never run before. If we wish at some future time to know whether in the 1950s there was enough computing power to solve the trajectory equations of a missile fast enough to act as a guidance system, the Pegasus simulation can be used to find out. Logical replication of this kind captures the persona of the machine and its behavior can be interrogated in the light of unforeseen enquiry.

One could argue that the computer running the simulation is itself subject to generational obsolescence and offers little to the custodial ambitions of museums to preserve machines in perpetuity. The simulation now runs on a Pentium-based platform, but what will run it in three years' time? With the promise of machine-independent software, there is the prospect of migrating the same simulation from one generational platform to the next and achieve, in the purity of logic, a kind of artifactual immortality—the dream of every conservator. So in the special case of electronic computers, logical simulation as a virtual object in some respects survives the forensic test of historical utility and suggests in prospect a salvation for the conservator for whom deterioration spells failure.[8]

I have narrowly defined the utility of original objects as their historical utility as evidentiary source. In this respect I would say that virtual objects do not pose a threat to the serious-minded business of forensic history for the same reason that physical replicas do not, however attractive and instructive replicas might be to the museum community and its publics. In the special case of computers, the virtual object—the simulation—agitates museologically intriguing ideas about artifactual identity as well as novel prospects for preservation. But if we look at broader social utility, then the situation is very different. The appeal of virtual objects as visitor attractions, their use as interpretative media, their opportunities for interactivity and education have substantial political appeal to managers and directors of museum-like institutions.[9] If the professional community that champions virtual objects competes with curators for attention, resources, and floor space, instead of seeing the formidable powers of new media as complementary vehicles to expound the value of the way in which objects and knowledge of objects inform understanding, then the threat is a very real one. For the notion of a museum as an entity inextricably linked to physical objects is one that begins to be challenged.

NOTES AND REFERENCES

1. Attempts to define the term "virtual object" are fraught with difficulty. "Virtual objects" as a category are philosophically problematic; the term is used to refer indiscriminately to a range of representations including holograms, still depiction using electronic photography, manipulable "digital surrogates," and fully interactive computer-based animations. I propose to bypass the obstructions to exact definition by

identifying the two characteristics cited, i.e., that the referent (the object) in the representation may be real or imagined, and that the medium of representation is electronic.

2. Noel Annan, "Gentleman vs. players." *New York Review of Books*, 29 September 1988, pp. 63–69.
3. Doron Swade, "Napoleon's waistcoat button: modern artifacts and museum culture." In *Museum collecting policies in modern science and technology* (London: Science Museum, 1991), pp. 19–26.
4. See, for example, Robert Anderson's discussion comment on p. 36 in: Otto Mayr, "Museum collections and the 'new technologies': the case of the Deutsches Museum." In *Museum collecting policies in modern science and technology* (London: Science Museum, 1991). Pierre Mounier-Kuhn, a French historian of computing technology has recently identified three instances in the history of computing in which reference to physical artifacts was the only course open in resolving fundamental questions not otherwise illuminated by any surviving documentary sources. Paper given at conference "Artifacts and History: Communications," Paris, 14–16 November 1999.
5. Doron Swade, *The Cogwheel Brain: Charles Babbage and the Quest to Build the First Computer* (London: Little, Brown, 2000).
6. Doron Swade, "Computer conservation and curatorship." *Resurrection: Bulletin of the Computer Conservation Society*, May, no. 1 (1990): 8–11.
7. For a dramatized version of Turing's speech, see Hugh Whitemore, *Breaking the Code*. (Oxford: Amber Lane Press, 1987), pp. 47–49.
8. The ultimate "virtual object" is computer software, which is by definition non-material. See Doron Swade, "Preserving Software in an Object-Centred Culture." In *History and Electronic Artifacts*, ed. E. Higgs (Oxford: Clarendon, 1998), pp. 195–206.
9. Doron Swade, "The Digital Superhighway and the Curator." In *Museum Collections and the Information Superhighway*, ed. Giskin Day (London: Science Museum, 1995), pp. 60–64.

Controversial Science Issues: Neither Apologetic Glorification nor "the Science War"

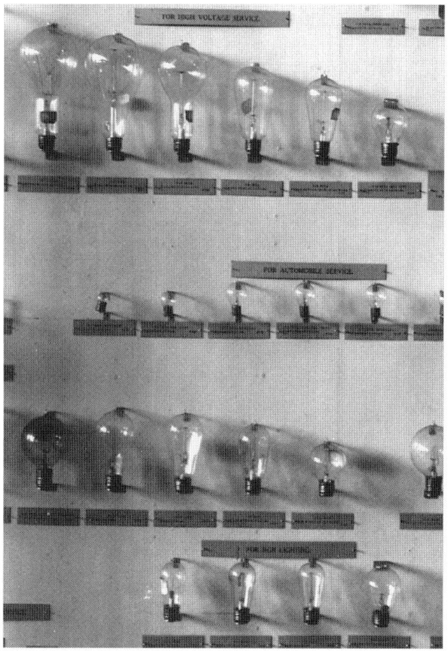

Context and Controversy

Bernard Finn

In the spring of 1963, during my first year as a fledgling curator at the Smithsonian, I made what was then for us a standard tour of European technical museums. Five weeks; ten countries; and some twenty museums. It was nice to be young and energetic and enthusiastic.

Out of this experience came two articles, one for *Museum News*,[1] the other for *Technology and Culture*.[2] In them I made a number of observations, tabulated comparative statistics, and pronounced my views on the state of the field. An unstated (but not unrecognizable) premise behind both articles was that there was some sort of Platonic "ideal" museum that all of us were trying (or should be trying) to emulate.

It didn't take much additional experience to realize that this was far from the case. Even if there were such an ideal, there would have been no agreement among museum professionals on what it looked like. But certainly such an ideal doesn't exist. We are too much the products of our own cultures, our own histories, collections, staff members. And for this very reason we have a lot to learn from each other—which is why it is still common practice (though not common enough in my opinion) for these visits to take place.[3]

What I would like to do today is describe how my own museum's circumstances (its historical context) have caused it to be different, and how that has led it into some interesting and controversial roles.

Most museums of science and technology (including, I believe, all of the other ones represented at this meeting) sprang at birth from a relatively narrowly conceived sense of scope and mission. Many grew directly or indirectly out of the great expositions of the nineteenth century,[4] some began with a core collection formed for some other purpose. But virtually all were founded independent of a pre-existing institution and were dedicated to the use of technical artifacts (which soon would become historical even if they weren't at the beginning) to educate the general public about aspects of science and technology. From these common origins they have developed, as I have already mentioned, in their own special directions. But the point I want to make here is that the Smithsonian was a different breed from the beginning, which helps to explain some of our peculiarities today.

We were born into a pre-existing institution, one in which, furthermore, the father figure didn't really want us. Joseph Henry, first Secretary of the Smithsonian and its guiding light for over three decades, was jealous of the limited resources he had at his disposal from the bequest ($500,000) that James Smithson had left to the United States government. Although the legislation establishing the Smithsonian in 1846 was purposely vague, and left open the possibility that it might engage in numerous activities—including those of a museum—Henry was rightfully wary of any operations that might detract from his primary purpose, which was to promote scientific research. But he also recognized that naturalists—who made up the bulk of America's scientific population—depended for their work on the things they collected and preserved from nature. And as early as 1848 he authorized the expenditure of a small amount of money to purchase such material and have it kept it in the building that was still being constructed. Two years later he appointed Spencer Baird, a naturalist, as his Assistant Secretary; Baird, needless to say, brought his personal—large—collection with him to Washington, effectively marking the beginning of a natural history museum at the Institution. Indeed, that same year the term "museum" begins to appear in Henry's annual reports to the Congress.[5]

Over the succeeding years occasional historical objects, including technical items, were added to the Smithsonian's collections. In 1857 the Congress voted to give the Smithsonian $4,000 per year to care for various collections that had come into government care—principally from exploring expeditions—and Henry was persuaded, however reluctantly, to take on a museum function (with Baird in charge). A dramatic increase occurred when the Smithsonian (meaning Baird, again) agreed to supervise the government's exhibits at the Centennial celebration in Philadelphia in 1876. As a "reward" the Institution was given all of the material exhibited, plus items from several foreign exhibits. These were then housed as a National Museum in a new Arts & Industries Building, paid for by the Congress, which opened its doors in 1881 next to the original building. Both still stand today. As can be seen from the floor plan, history—meaning both social and technical history—had arrived. (Figure 1) But they had obviously arrived as poor second cousins to a dominant natural history. This can also be seen in the expanded curatorial staff in the 1890s.

There were two consequences of this close proximity of history and natural history. One was a taxonomic approach to history, with a strong emphasis on the value of the objects. The other was a progressive evolutionary organization applied to exhibits. G. Brown Goode—Baird's assistant, who took over responsibility for the museum after Baird succeeded Henry as Secretary in 1879—described as an ideal exhibit one that showed, for instance, how a useful device emerged from the successful manipulation of various materials.[6]

In 1911 Natural History fled across the Mall to a new building, leaving the historical collections an opportunity to expand into the vacated space. (Figure 2) Which they rapidly did. But this was a collection of collections, one might argue, and not a purposefully constructed museum enterprise. One thing to be noted is that the evolutionary themes, which suggested some form of context, generally gave way to displays of simple progress or commemoration. (Figure 3 and Figure 4)

Between the wars, when Smithsonian funding was low and initiatives generally lacking, a small group within the museum began to agitate for a museum of science and industry.[7] By now there were impressive models in several European countries—most notably in England, France, and Germany. These efforts gave rise after the Second World War to a Congressional appropriation for a new museum building. But the lack of any clear tradition—or perhaps the existence of a rather muddied tradition—led in the end to a decision not to leave social and cultural history behind.[8] The building that opened its doors in 1964 therefore became the Museum of History and Technology. A separate technical museum came a dozen years later, in the form of the National Air and Space Museum—and the immediate impetus for its creation rose not from within the Smithsonian but from Congress itself, encouraged by NASA and the air and space industries.[9]

The establishment of a new museum embracing the history of science and technology and social and cultural history was significant because it was a positive step, not just the remnants of what was left over from someone else's move. But even in the phrasing of the new name—Museum of History and Technology—there was an implication that there were actually two museums with no particular expectations of interaction. The organization further reinforced this view. The structure was along traditional lines. Subject areas—whether physics or medicine or politics or costume—were

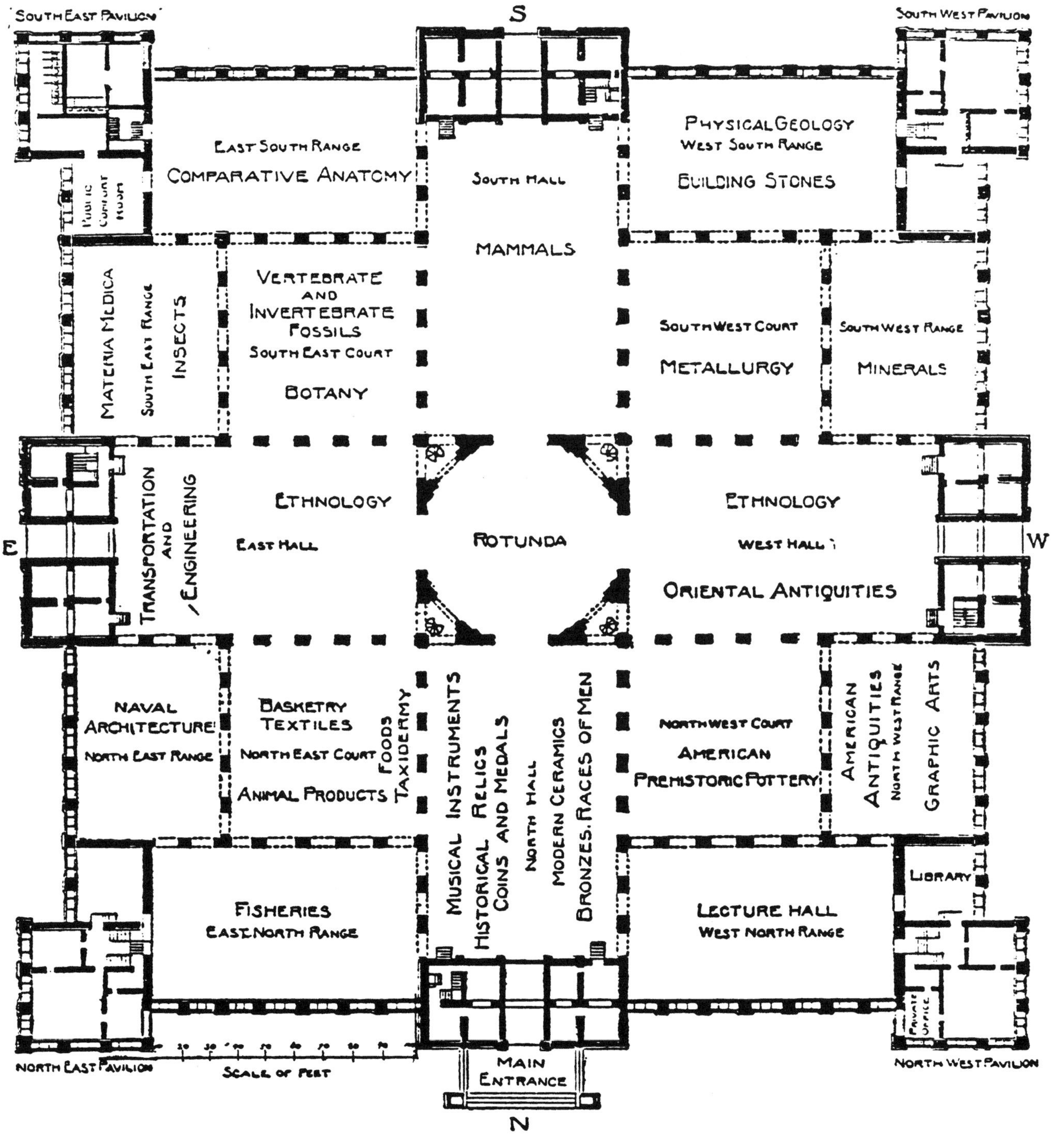

FIGURE 1. Floor Plan of Arts & Industries Building, 1888. From Report of the U.S. National Museum, 1888 (Washington: Government Printing Office, 1990), p. 26.

narrowly conceived, based on collections. And museum structure followed: electrical collections, electrical curator, electrical display space. Nevertheless, proximity of the two groups inevitably fostered some degree of mutual interest, understanding, and even respect.

But there was another factor, even more important in my opinion. At the time when new staff were being hired for the new museum, in the late 1950s and early 1960s, academically trained historians of science and technology were beginning to be available. A sufficient number of these were

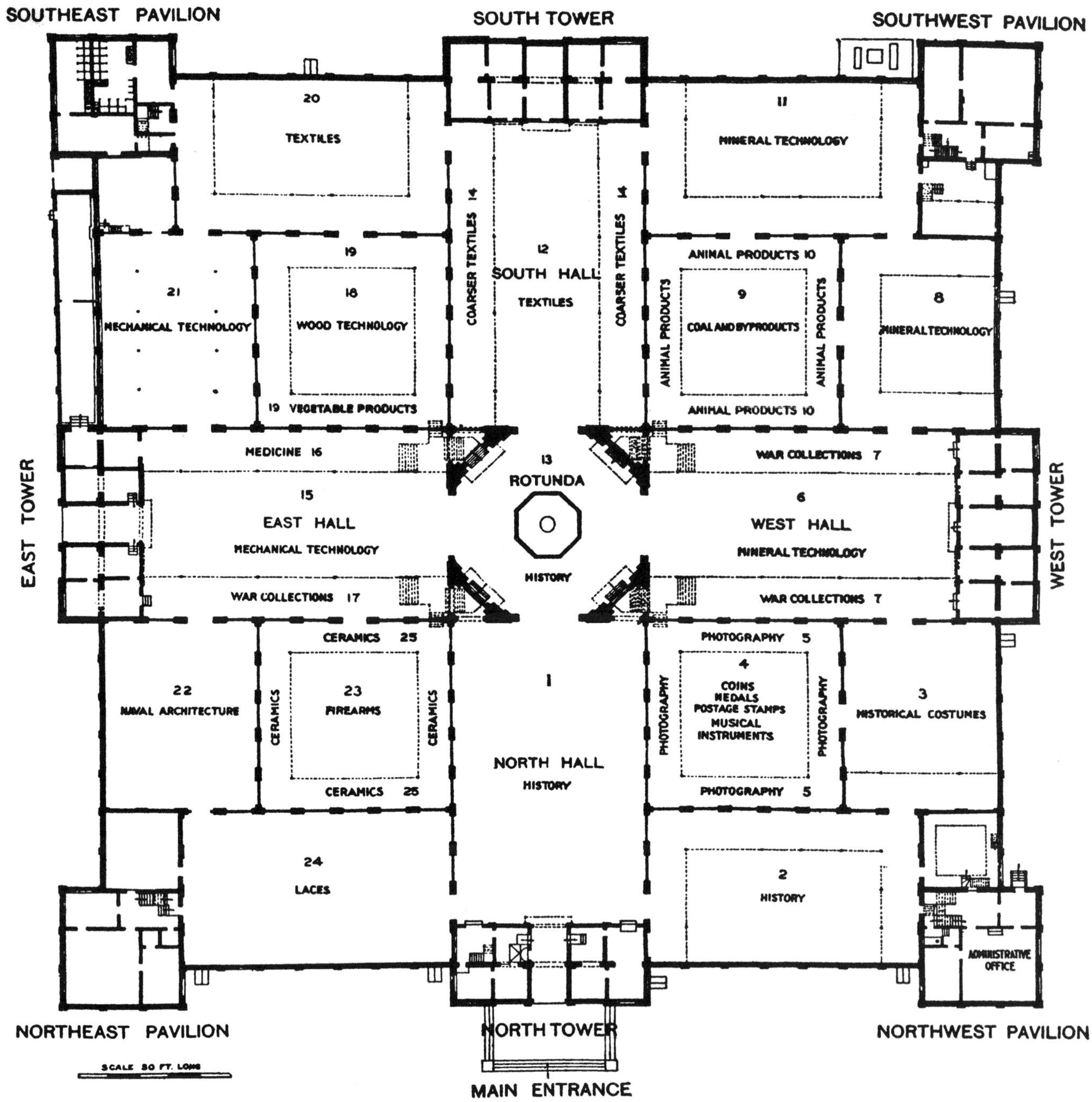

FIGURE 2. Floor Plan of Arts & Industries Building, 1919. From Report of the U.S. National Museum, 1919 (Washington: Government Printing Office, 1920), plate 7.

hired to set the tone for the museum.[10] Furthermore, these historians maintained their affiliations with their academically-oriented societies.[11] This meant that as the profession changed and became more contextually, socially conscious, so did they. The consequence of these two factors—a museum that included social and cultural history, and historians as curators—meant that by the mid-1970s exhibits were no longer conceived in whiggish, internal terms; the technical material was increasingly placed in a story-telling mode, with social themes and context.[12] This set the stage for a new name for the museum in 1981—the National Museum of American History, which implies a degree of integration substantially greater than actually exists.

There have been gains and losses in this transition. The greatest gain has been a closer tie between

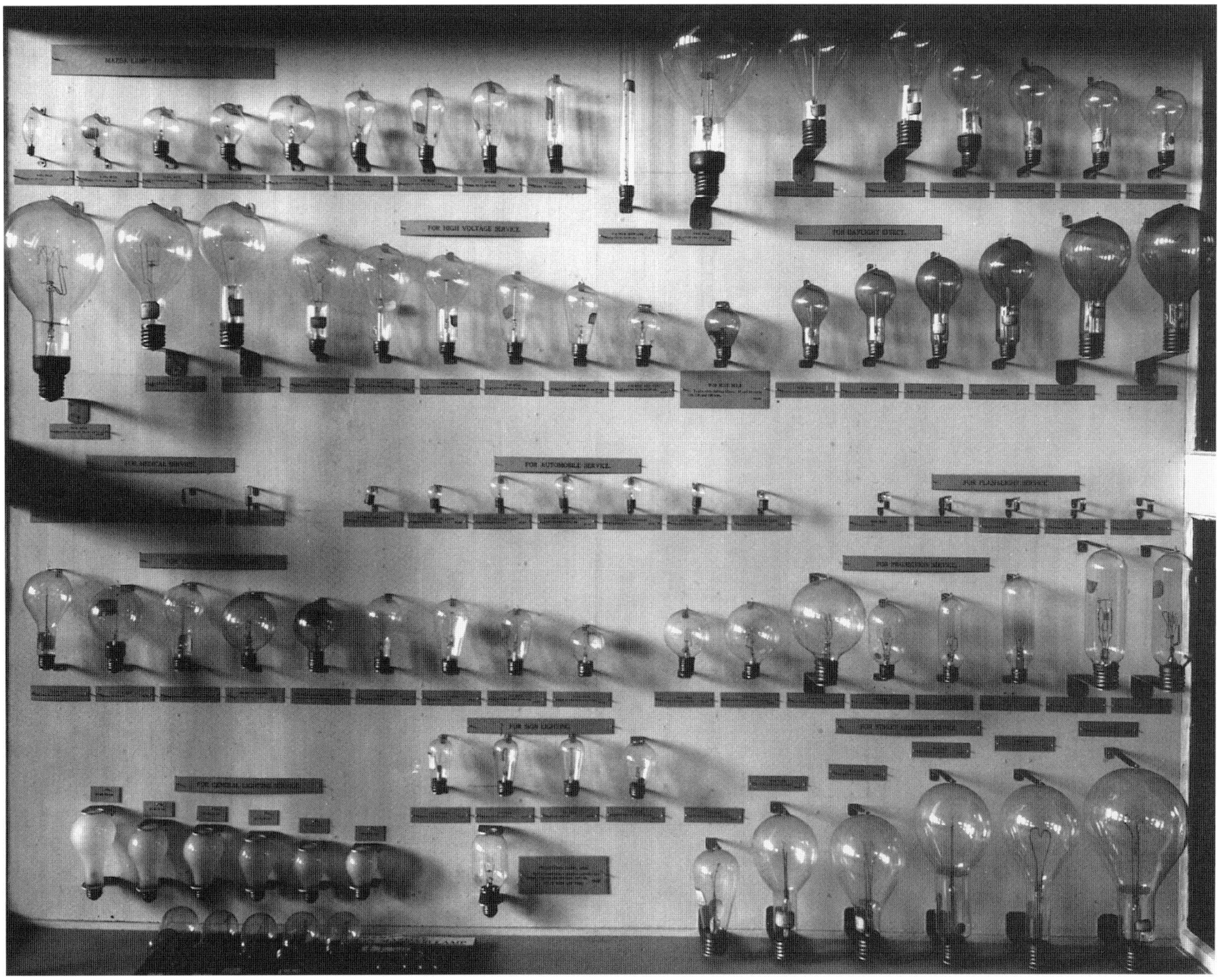

FIGURE 3. Portion of Electric Light Exhibit in A&I Building, 1930s.
Courtesy Smithsonian Institution.

museum displays and recent historical scholarship; and the story-telling mode is presumably more appealing to visitors, especially to those who have no particular interest in the subject matter. The greatest losses have been in the use of artifacts, which are increasingly used as symbols rather than prime source material; and the story-telling mode may actually have a negative influence on those who aren't interested in that particular theme.

There have also been consequences.

There has been a shift in arrogance.

Museums, as public institutions, have traditionally been viewed like encyclopedias, as sources of information, as objective authorities. Earlier that authority, which often could be perceived as arrogance, concerned objects—what they were, why they were important, how they changed, how they worked. There were pitfalls in this approach. The museum was seen (and rightly so) to be speaking from the point of view of the scientific and engineering community. Even if science and technology occasionally caused problems, they eventually solved those and other problems. And the progressiveness of these fields was easily translated into a tone of social progressivism. This was fine for the museum's principal supporters—scientists and engineers. It increasingly annoyed academic historians, but their voice was muted. An example of this was the review of the Air and Space Museum written by Michal McMahon and published in *Technology and Culture* in 1977.[13] He strongly criticized the museum for being a publicity front for the air and space

FIGURE 4. Cyrus Field Commemorative Exhibit in A&I Building, 1930s.
Courtesy Smithsonian Institution.

industries and for ignoring the destructive uses of these devices. He called on the museum to display the *Enola Gay*, which had dropped the atomic bomb on Hiroshima, as the ultimate example of the power these devices controlled. The plea fell on deaf ears, at least in terms of policy implications.

The new mode of the museum might be called an arrogance of ideas. The authority of the museum has been placed on the stories being told, on the context into which objects are being placed.

One consequence of this is the development of a much more partisan audience. Those who agree

with the ideas applaud, those who disagree write letters to the editor. It is not a trivial fact that those who are most likely to be offended are those who previously were the museum's staunchest supporters. The scientists and engineers generally like to have their professions portrayed in a positive, nonthreatening way. They are not as pleased to have noted some of the collateral effects—such as water pollution or smog or bombs. And when we finally did get around to displaying the *Enola Gay*, it was people from this group that became most upset.[14]

A second consequence is the loss of respect for the individual object. Instead of being shown in their technical context, where one can see and perhaps understand their changing functions, they are more often shown as symbolic representations in some overarching social story.

Is there some compromise? Can we retain some measure of authority and still deal with historically valuable themes? Can we use the objects in thoughtful and provocative ways?

The answer to all of these questions is yes, but not easily. I think we need to encourage visitors to question our premises, not to accept our authority blindly. This can be done by providing alternative views. Or by "signing" exhibits with the names of individuals (not the anonymous museum) responsible. Both of these have been tried, with mixed results. More needs to be done.

Stories need to be written around the objects, looking to them for evidence of points to be made. Much more often the story comes first (from the academics) and the objects are hung on as ornaments. We need to force ourselves to do otherwise.

NOTES

1. Bernard Finn, "The New Technical Museums," *Museum News* 43 (Nov. 1964): 22–26.
2. Bernard Finn, "The Science Museum Today," *Technology and Culture* 4 (1965): 74–81.
3. It is commonplace for planning committees to tour existing museums, a practice at least a hundred years old—as chronicled in Warren Perry, *The Science Museum of Victoria: A History of Its First Hundred Years* (Melbourne: Science Museum of Victoria, 1972).
4. Eugene Ferguson, "Technical Museums and International Exhibitions," *Technology & Culture* 6 (1965): 30–46.
5. William Washburn, "Joseph Henry's Conception of the Purpose of the Smithsonian Institution," in *A Cabinet of Curiosities* (University of Virginia Press, 1967), pp. 106–166.
6. For a discussion of the Smithsonian's exhibition policy in this period, see Sally Kohlstedt, *The Origins of Natural Science in America: The Essays of George Brown Goode* (Washington: Smithsonian Institution Press, 1991).
7. Arthur Molella, "The Museum that Might Have Been: The Smithsonian's National Museum of Engineering and Industry," *Technology & Culture* 32 (1991): 237–263.
8. Frank Taylor, the principal proponent of the museum, recalled that the decision was made in casual, almost off-handed fashion in the early stages of preparation of a submission to Congress, probably in 1950 or 1951. As justification he noted that, for instance, the military history people were interested in various forms of technology. Frank A. Taylor Oral History Interview 4, 1974, Smithsonian Institution Archives, RU 9512, p. 94.
9. The National Air Museum was created as part of the Smithsonian in 1946, but it never received sufficient funding for a major facility. In 1964 Congress authorized planning money for a new museum building and in 1966 changed the name to the National Air and Space Museum. But little was achieved until members of Congress exerted further pressure on that institution and on the Smithsonian in the early seventies. See, in particular, Alex Roland, "Celebration or Education? The Goals of the U.S. National Air and Space Museum," in *Industrial Societies and Its Museums*, ed. Brigitte Schroeder-Gudehus, special issue of *History and Technology* 10, nos. 1&2 (1993): 77–90.
10. In the early 1950s, before authorization of the new museum, the staff consisted of about a dozen curators, only a few with training in history and none with advanced degrees. In 1964, when MHT was inaugurated, in a staff of 42 curators almost half had graduate training in history, including twelve with doctoral degrees.
11. John Schlebecker was president of the Agricultural History Society, Brooke Hindle served as president of the Society for the History of Technology, and Robert Multhauf was president of both SHOT and the History of Science Society. Other curators have been on committees of these and other historical societies. *Isis* and *Technology and Culture* have both had editorial offices at the museum, *Isis* for fifteen years under Multhauf, *Technology and Culture* for eight years under Robert Post.

12. There were four major exhibits in particular: *Person to Person* (telegraphy and telephony) in 1876, *Atom Smashers* (particle accelerators) in 1977, *Hall of American Maritime Enterprise* in 1978, and *Edison: Lighting a Revolution* in 1979.
13. Michal McMahon, "The Romance of Technological Progress: A Critical Review of the National Air and Space Museum," in *Technology and Culture* 22 (1981). An example of concern from outside the academic community is Howard Learner, *White Paper on Science Museums* (Washington: Center for Science in the Public Interest, 1979).
14. The script for the proposed exhibit placed the bombing in a broad context that included the previous history of strategic bombing, the subsequent history of the Cold War, racial feelings in the United States, and cultural feelings in Japan. When it was made public in the spring of 1994, more than a year before the proposed opening, the reaction of the Air Force Association was vehemently negative and politically persuasive. The exhibit was replaced by one which confined itself to an expression of the value of this technology in ending the war. Among other sources, see *Judgement at the Smithsonian*, ed. Philip Nobile (New York: Marlowe & Company, 1995); also Michael Wallace, "The Battle of the Enola Gay," in his *Mickey Mouse History and other Essays on American Memory* (Philadelphia: Temple University Press, 1996), pp. 269–318.

The Power of the Pencil

RENEGOTIATING THE MUSEUM-VISITOR RELATIONSHIP THROUGH DISCUSSION EXHIBITS

Ben Gammon
Xerxes Mazda

INTRODUCTION

Visiting an exhibition about science and technology is generally a process of one-way communication—the museum speaking to the visitor. In contrast, the idea of allowing visitors to express and record their opinions is a new and ground-breaking departure, at least for museums of science. To include visitor comments in the exhibition and to give them status equivalent to the material produced by the museum is an even greater break with tradition.

An exhibit that enables visitors to express their opinions would seem to hold much potential as a novel interpretative tool. It involves the visitors and draws them into complex discussion far more successfully than static text, objects, or even interactive exhibits. This is useful because displays about modern science increasingly attempt to cover complex, unresolved issues. These topics do not easily lend themselves to the use of traditional exhibition media.

Although the use of such visitor feedback exhibits is on the increase,[1] there has been surprisingly little assessment of how well they fulfill the objectives of either the museum or the visitors. Only a handful of studies have been published that explore the background and the effectiveness of this method of interpretation.[2]

This paper is concerned with an analysis of one method of garnering visitor feedback—discussion exhibits.[3] It begins by defining a discussion exhibit and describing the experimental use of this kind of exhibit in the Science Museum, London. The next section critically examines the motivations of the Museum in providing discussion exhibits and the motivations of the visitor when using them. From this analysis we suggest that, although discussion exhibits can be effective and engaging, there are three possible pitfalls which may compromise their success.

WHAT IS A DISCUSSION EXHIBIT?

A discussion exhibit is defined as an area where visitors can write their opinions or questions about issues covered by the surrounding exhibition. Visitors are prompted by a series of open-ended questions, for example, "What do you think about the Big Bang as a theory of the origin of the universe?" The visitors write comments on cards or sheets of paper and post their responses into a ballot box. Museum staff periodically empty the box and screen the comments to remove those that are irrelevant, obscene or potentially libellous. The remaining

comments are then typed, laminated and added to the display of previous comments left by visitors. These discussion exhibits are never free-standing, but always form part of larger exhibitions. This definition of discussion exhibits excludes examples in which visitors are asked to select answers from a pre-determined list or to vote on a particular issue. Nor does this definition include comments books where visitors are asked to comment on their visit to the museum.

DISCUSSION EXHIBITS AT THE SCIENCE MUSEUM, LONDON

For several years the Science Museum has been using discussion exhibits with varying success.[4] In order to better understand the possible pitfalls of such exhibits, the Museum recently began a program of research into their use. Discussion exhibits were developed as part of a series of temporary exhibitions. Each exhibition covered a contemporary issue in science, technology or medicine and was aimed at a non-specialist audience aged 12 and above. The discussion exhibits were evaluated to determine how visitors were reacting to them and to assess what the project teams were hoping to achieve by including them in the exhibition.

This paper will focus on discussion exhibits from the following three exhibitions:

1. *Future Foods?* looked at the science and issues behind genetically modified food.
2. *Join the Great Fat Debate* aimed to introduce visitors to the issues around Olestra—a manufactured substitute for fat.
3. *The Big Bang* covered the theory of the Big Bang, and the research that led to that theory.[5]

WHAT MOTIVATES A MUSEUM TO USE DISCUSSION EXHIBITS?

The second part of this paper explores the exhibit developers' expectations of the discussion exhibits. This was researched through interviewing members of the project teams for each exhibition, and analyzing internal Museum documents. The research found that Museum staff believe discussion exhibits encourage visitors to engage with what is perceived to be difficult material.

An exhibit developer has available a range of techniques for presenting science to the public. However, it is extremely difficult to cover predominantly issues-based subjects using traditional exhibition media. Many issues in contemporary science are abstract and do not easily lend themselves to the use of interactives nor to more traditional object-based displays. There is a danger that such exhibitions can become text heavy, with little appeal for the visitor. Previous experience of trying to develop issues-based exhibitions proved to be highly unpopular with visitors. For example, an exhibition about the issues behind screening for genetic illnesses was described by visitors to be "uninspiring," and "drab and dreary."[6] As one visitor summarized: "If you had children with you then you would [walk by] because you couldn't keep them still while you sat and read. There's nothing to occupy them."[7]

The use of discussion exhibits is often seen as a way of getting around these problems. For example, the *Future Foods?* project team felt that "previous exhibitions have shown that a comments box is a successful way of allowing visitors the chance to explore their feelings about a variety of issues."[8]

Similarly, *Join the Great Fat Debate* was developed "so visitors could directly participate in an exhibition about a chemical subject," because the exhibition developers felt that "it is not only scientists who have a valid opinion about science and technology."[9]

The developers of the *Big Bang* exhibition were explicit in their belief that "The Big Bang theory raised contentious issues and some people would hold quite strong beliefs that they would want to be able to communicate . . . So we hoped that having a feedback book would prevent visitors from feeling any frustration at not seeing their personal views represented in the exhibition."[10]

Evaluation of the discussion exhibits confirmed that the project team's wishes were met. For example, observation studies showed that 52% of the visitors to *Join the Great Fat Debate* used the discussion exhibit by reading and/or writing comments.[11] When visitors to this exhibition were interviewed about their reaction to the idea of developing more

FIGURE 1. "Vote with your trolley"—the discussion exhibit in the *Future Foods?* exhibition at the Science Museum.
Science Museum/Science and Society Picture Library

discussion exhibits, there was an overwhelmingly positive response.

> "Go for it—it would give the public perspective."[12]
>
> "People should be prompted to give an opinion rather than go along with it without thinking."[13]

A series of focus groups and in-depth interviews were conducted to assess visitors' reactions to the *Future Foods?* exhibition. On the whole visitors responded very favorably to the discussion exhibit. As stated in the final report, "The visitors' comments book was seen in a very positive light. In fact for many it provided an important role in providing balance."[14]

Some of the most convincing evidence for the effectiveness of discussion exhibits is in the number and quality of responses that were left by visitors. Over a three-month period, more than 2,000 comments were left in each of the three exhibitions. In many cases the quality and length of these comments illustrated the care and time which

FIGURE 2. The discussion exhibit in *Join the Great Fat Debate* at the Science Museum.

Science Museum/Science and Society Picture Library

visitors were giving to their responses. The following two examples are typical of many comments that were left.[15]

"I simply don't trust scientists (sorry!) to determine the future composition of my food. There is increasing evidence to link between certain diseases, notably cancer, and sprays, insecticides and other chemicals. At present we simply do not know the long-term genetic (human) implications of genetic tampering. I should vastly prefer my food to be as nature engineered it, spots and all!—at least then I know and can be responsible for what I eat, rather than having other people engineer it for me. The future of our food is terrifying, and particularly the fact that we have now reached a stage where whatever we buy, we cannot be sure whether or not it has been interfered with."[16]

"There's enough fresh fruit and vegetables on the market. Adding extra artificial foods can only bring long-term problems to society's diets. I feel that the money that has been spent on researching such 'new' foods is better spent on promoting a healthy, natural diet, with a moderate amount of fat."[17]

Similar results have been found for discussion exhibits in other museums. For example, the *Share your reaction* exhibits in the Art Gallery of Ontario received around 5,000 responses over a nine-month period. The quality of these comments was also deemed to be very high:

"The range of responses have often left me quite speechless because of their power and mystery and none of them really reflects the kind of insight into the art experience that the gallery itself could articulate."[18]

Although the desire to have an effective interpretative tool was found to be the major motivation for exhibition teams, the Science Museum was also keen to explore the use of discussion exhibits as a way of catalyzing debate on galleries without the use of expensive staff moderation. This is part of the Science Museum's aim to become a center for public debate on issues in contemporary science and technology.

It is interesting to ask how far discussion exhibits can fulfill this role. An analysis of the responses left by visitors showed that real debates were emerging. The following are four consecutive responses left in the *Big Bang* discussion exhibit. In this sequence, visitors refer back to previous comments, which gives the impression of a dialogue.

"A theory is just a theory. We cannot be sure if it is reality. As long as there are no experiences made, that speak against it, it has to be taken for true. To me, the Big Bang appears to be a quite good explanation. I hope it is true. If it is, we have done one more step to wisdom. But no-one will ever know!"

"The validity of this highly contentious theory is proved by the exhibit—simplistic, naive, lacking in proofs. Just look at the rest of the museum for inspiration! Allah, creation!"

"And the Koran has proof in it does it? That's a big NO."

"But this exhibition SAYS Big Bang is a theory! And it says about the 'proofs' to back it up. For example the background radiation and the amount of helium. I don't think it's a contentious theory at all, it seems to have more evidence than steady state or Creationist views of the Universe. I know what I'd rather believe."[19]

In addition, the *Future Foods?* focus group displayed considerable support for the idea of extending the discussion element by having feedback from experts as well as visitors. Comments to this effect were even written by visitors to the exhibition: "It would be a great idea to answer the questions in this exposition."[20] Although the practicalities of regular expert response to visitors' comments are awesome, it is something that the Museum will attempt to do in the future.

VISITORS' MOTIVATION

The third part of this paper looks at visitors' motivations to take part in discussion exhibits—both to read and to write comments. Three key motives were found for visitors to write comments and three for visitors to read comments left by other people.

Motive to write 1: An outlet for anger

Visitors seemed to be using the discussion exhibit as an outlet for their anger and frustration. Many of the comments showed real passion, with words written in capitals or heavily underlined. The specific points raised in the exhibition appeared to have acted as a catalyst for visitors to express opinions drawn from their own knowledge, experience and prejudices. The following two examples illustrate this.

"Where is the democracy?! It is the consumers who do not benefit. The companies who produce it are there to make a profit! All genetically engineered food should be banned or at the very least labelled in shops so we are not hoodwinked into buying it."[21]

"Olestra adds no calories to the diet. It cannot be used by the body. How much has it cost to develop and how much profit do the manufacturers want to get back from it? Twenty-five years of testing is an awful lot of investment. Why do we need it? Answer: we don't, but someone thinks they can profit from it £$."[22]

Motive to write 2: A wish to cause change

Visitors seem to be motivated by the belief that their comments would be read by manufacturers and policy makers. This belief is not particularly surprising, because almost the only time public opinion is sought is in market research. A *Future Foods?* focus group participant voiced the belief of many visitors when she said:

"I think it is a very controversial topic and a lot of people are very upset about it. It stirs emotions. So I think they're trying to find out what people think, and then they can use that to focus their marketing strategy to counteract, because after all they're selling a product of some sort."[23]

Motive to write 3: The graffiti effect

The desire to scribble graffiti appears to be a powerful motivation for visitors to use the discussion exhibit. Although many of the examples shown so far have been of a high quality, it would be wrong to imply that this was always the case. On the contrary, many of the comments are best described as graffiti. Of the 2,259 comments left in the *Future Foods?* exhibition, only 34% were deemed relevant to the exhibition,[24] while for *Join the Great Fat Debate*, the percentage was 22%.[25] The largest proportion of "graffiti" comments comprised scribbles, with a smaller amount of obscenities and general views about the Museum.

Many of the comments that were considered relevant and were displayed in the exhibition comprised very short sentences, with only the most tenuous link to the question that was asked. The following examples, taken from *Join the Great Fat Debate*, in response to the question "What do you think about olestra?" amply illustrate this point:

"Eat it if it tastes nice."

"Olestra is super-fat man."

"What is olestra?"

"Gemma likes it."

"Chocolate is my favourite food."[26]

MOTIVATION TO READ

It was also possible to identify three motives for visitors to read the comments left by previous visitors.

Motive to read 1: language

The first motive to read comments concerns the style and language used by visitors when expressing themselves. The comments make compelling reading. Why is this so? One possible explanation lies in the emotive language that visitors use, as illustrated by these examples:

"I am a molecular biologist and GM food scares me to death."

"No problem with genetic mods or clones. If we had let religion have its way we'd still be in the caves."

"GE is not an extension of existing selective breading. GE is clumsy and dirty. In some processes tiny particles of tungsten are fired into the chromosomes of 'host cells [........] its like firing a cannonball at a butterfly with a maggot and an appleseed attached and hoping it sticks in the eyeball and not the wing! Stop it you silly people."[27]

The visitor who wrote this third comment certainly had a good working knowledge of genetic engineering, and yet, unlike explanations written by the Museum, the style of writing is very emotive, expressive and ultimately very readable. It is quite conceivable that many of the points made by this visitor could have been made by the Science Museum, but the language would be more measured, balanced and neutral.

Motive to read 2: balance

The second reason why visitors read the comments left by other visitors is concerned with bias. Visitors felt the *Future Foods?* exhibition was biased in favor of genetically modified food.[28] For example:

> "This exhibition boldly states that risk assessments examine all potential effects that GM could have on our health and environment then goes on to list impressive sounding committees. This is a ridiculous and scientifically inaccurate claim. It is impossible to assess how a gene pool will behave—just as nobody foresaw the impact of introducing new species in Australia or New Zealand."[29]

The book of comments was valued by visitors as an essential component of the exhibition, as it was perceived as a method of redressing the imbalance. As one focus group participant commented, "It was interesting that in the comments book it was all the opposite way, so maybe there was some balance there."[30] Visitors even wrote comments to this effect in the book: "This seems to me to be a very agribusiness dominated exhibition. The only doubts are ours, the punters, on these slips of paper."[31]

Motive to read 3: a novel point of view

Finally, the discussion exhibits were valued by visitors as they allowed issues to be tackled from the point of view of the visitor rather than that of a scientist, exhibit developer or manufacturer.

To summarize, museums see discussion exhibits as valuable components of exhibitions because they provide effective methods of interpretation which involve visitors, encouraging them to engage with the complex issues in modern science.

Visitors, on the other hand, are motivated to write their comments by a desire to vent their anger and register their concern over an aspect of the exhibition material. Some visitors may take this further, believing that their comments could affect the technology under discussion, and yet other visitors respond to the same impulse that drives them to add to graffiti in a public place. The resulting collection of visitor comments makes compelling reading owing to its emotive language, the way it addresses visitors' concerns rather than those of the exhibit developer, and the way it can make up for perceived bias in the exhibition.

So, on the surface, there appears to be a good match between the Museum's primary expectations and the visitors' experiences. The Museum wants the visitor to engage with complex ideas expressed in the exhibition, and the visitor certainly finds the discussion exhibit engaging. Yet it is necessary to sound a note of caution. This research has identified some potential problems which need to be addressed. We suggest that both the Museum and the visitor need to be aware of the following three potential delusions.

THE DELUSION OF UNIVERSAL APPLICATION

It is tempting for exhibit developers to believe that discussion exhibits could work for all subjects. Yet having teased out the visitors' motivations, it becomes clear that this is not the case. Discussion exhibits work best on subjects that visitors feel strongly about—those covering controversial and emotive issues. The more detached the visitor feels from an issue, the weaker become the anger and the desire to effect change, and there is a danger that the primary visitor motive to contribute becomes the graffiti effect. By its nature this would make the written comments less compelling to read, thereby downgrading the degree of visitor interaction with the discussion exhibit as a whole. This has indeed been the case in other discussion exhibits tried at the Science Museum. For example, in the exhibition *Bucky Balls*, a discussion exhibit was included which posed the question "What do you think Bucky Balls might be used for in the future?"[32] There was no compelling motivation for visitors to answer this question, as evidenced by the fact that less than 20 comments were left over the five months the exhibition was open.

THE DELUSION OF MARKET RESEARCH

To all intents and purposes, the discussion exhibit looks to a visitor like a tool of market research. This paper has already presented evidence that the visitor believes the data is being collected by the Museum on behalf of manufacturers.

This illusion has several ramifications. Firstly, the Museum is in danger of compromising its neutrality. How can visitors believe the Museum is neutral if they see that the Museum is acting as an agent for manufacturers by acquiring commercially valuable market information?

Secondly, there is a danger that the visitor is being deceived. If visitors believe that they are indirectly shaping technologies because the people who can make a difference will read their comments, then what will these same visitors think if they find out that their comments are instead kept in a file in the museum archives? This probably over-exaggerates the case. It would certainly be very difficult for a museum to hand comments over to a manufacturer without compromising the museum's position of neutrality. However, during the course of an exhibition the comments are displayed for all to see. Both the manufacturer and groups representing other sides of the debate are likely to visit the museum in order to investigate the range of concerns that people would have.

THE DELUSION OF LAY DISCUSSION

Who contributes to discussion exhibits? Is it the lay public, people with a specialist knowledge, or representatives of the manufacturers or pressure groups? It is probably all of these, but importantly there is no way of telling, as all comments are anonymous. Without evidence to the contrary, museums and their visitors presume that the comments are those of the general public. What is to stop manufacturers or pressure groups from anonymously adding comments in an organized fashion? Admittedly it seems a lot of effort to go to when discussion exhibits currently have such a low profile and it is unlikely that either *Future Foods?* or *Join the Great Fat Debate* were targeted in this fashion. However, if ever discussion exhibits gain a higher profile, they are likely to be highjacked by organizations with financial or moral interests in the subject matter, and both the museum and visitors will be deluded.

CONCLUSIONS

The above research has convinced us that discussion exhibits are a popular, cheap and effective interpretative tool for engaging visitors in debates about emotional and controversial scientific issues. Using the findings of the research, a second generation of these exhibits is being developed for use in the galleries of the new Wellcome Wing at the Science Museum, London. Taking a wider view, there is no doubt that discussion exhibits will increasingly be used in museums and galleries, whether they are concerned with science or the arts.

The work outlined above has exposed some of the ground rules and pitfalls of discussion exhibits. As these exhibits increasingly play a useful role in museums of modern science, project teams should keep a firm eye on the three delusions in an attempt to minimize their impact. With this caveat, we believe discussion exhibits have a bright future in renegotiating the traditional relationship between museums and their visitors.

NOTES

1. The following examples of feedback exhibits show the range of techniques used:
   - The Jewish Children's Museum in Los Angeles, California has a replica of the Wailing Wall in Jerusalem. Children are provided with paper to write their comments, which could then be stuck to the replica wall with tape. The most interesting and poignant notes are displayed for other children to read.
   - The Sixth Floor Museum at Dealey Plaza in Dallas, Texas has a Memory Book area where visitors can record whatever they want after visiting the site where the evidence was found after John F. Kennedy's assassination.
   - The Tech Museum of Innovation in San Jose, California has two discussion exhibits. *Innovation Forum* presents audio-visual statements about technology and innovation by a number of leaders in various fields. Visitors are encouraged to record their own thoughts, which are made available for display as part of the exhibit. The *Television V-Chip* introduces the issue of controlling the impact of "bad" TV content on the public. Visitors record their opinions on paper about the V-Chip and use of television. Thoughtful comments are saved in a ring binder for other visitors to read.
   - At the Monterey Bay Aquarium, visitors use index cards to record their opinions about ocean conservation. Many of these are then displayed on the walls.

- Visitors to the Bradbury Science Museum in Los Alamos, New Mexico, have filled 18 ledger books over a 6-year period with comments about an exhibit on alternative perspectives on nuclear weapons.
- At the U.S. Holocaust Museum in Washington, D.C., there is a book for visitors to write their thoughts after seeing the exhibition.
- The Denver Museum of Natural History in Colorado included discussion exhibits in their exhibition *Edge of the Wild*. Visitors were invited to write about their encounters with wildlife. Museum personnel chose some of these comments to display in the exhibition.
- A comments book was included in an exhibit on sweatshops at the National Museum of American History, in Washington, D.C.
- When the Art Gallery of Ontario in Toronto redeveloped the galleries displaying the Canadian Historical Collection, they included 24 discussion exhibits called *Share your reaction*, where visitors were encouraged to relate the personal associations they had with the paintings on display. Visitors could draw and/or write comments on cards which they posted into boxes. This was further developed for the *OH! Canada* project, where visitors could write or draw their responses to questions about the Canadian environment and Canadian identity. Specific questions were displayed each week in the exhibition and in newspapers and on the Internet. Various forms of feedback were used—response cards, graffiti wall (a gigantic chalk board), video feedback and feedback via the Internet and by fax machine.
- The Walsall Museum and Art Gallery, UK has used a range of different feedback exhibits over the past few years to invite visitors to write, draw or paint in response to their exhibitions. For example, children could take home stamped-addressed postcards to draw what they remembered of the exhibition. These were then sent back to the museum. Visitors were also asked to vote on their most and least favorite paintings in the collections, explaining why. Some of the best comments were used as label text for the paintings.
- In the *Not just homeless* exhibition at The Pump House: People's History Museum in Manchester, UK visitors could write their comments on post-it notes and stick them to a wall.
- In *The Museum of Me*, London, each visitor adds to a growing archive of personal details—they may write about themselves on paper, blackboards and walls, pick a favorite fabric, mark their location with a pin on a map, or measure their passion levels on a fairground machine.

2. For example, see Douglas Worts, "Gallery enhancement: Forging a new partnership with the public," in *Visitor Studies: Theory, Research and Practice 6* (Jacksonville, Alabama: Visitor Studies Association, 1993), pp. 176–197.
3. Other forms of visitor feedback would include comments books, video stations, audio recording, public debates, gallery drama, consensus conferences, and various methods of voting.
4. An early example of feedback in the Science Museum, London, was mentioned in the Annual Report for 1937. A temporary exhibition about electric illumination was attended by about a quarter of a million people over five months. The report states that "by means of an operable exhibit, 147,272 persons recorded opinions about the amount of light desirable for comfortable reading and a large number recorded their decisions about a heterochromatic photometry experiment. The Museum had been asked to preserve the results of these tests as a basis from which important conclusions were to be drawn—an example of how the Science Museum can fulfill a useful function where mass statistics are required as the material of scientific investigation. The co-operation of the public was marked by the care and intelligent interest which the majority displayed."
5. The *Big Bang* discussion exhibit differed from other discussion exhibits in that visitors wrote their contributions on a sheet of paper bound into the same flip-book as the typed-up comments of previous days. It was therefore much more like a traditional comments book.
6. *Genetic Choices?* was a temporary exhibition about the issues behind screening for inherited diseases.
7. *Genetic Choices?—Results of a visitor evaluation*, an unpublished internal Science Museum report produced by Creative Research Ltd. (1997).
8. *Future Foods?—Visitor Comments*, an unpublished internal Science Museum report, ed. Nicola Perrin (1998).
9. Xerxes Mazda, Yvonne Harris and Ben Gammon, *Join the Great Fat Debate: The production of a discussion exhibit at the Science Museum, London*, an unpublished internal Science Museum report (1998).
10. Sarah Hunt, personal communication (1998).
11. Yvonne Harris and Ben Gammon, *Join the Great Fat Debate summative evaluation*, an unpublished internal Science Museum report (1998).

12. Ibid.
13. Ibid.
14. *Evaluation of Future Foods?*, an unpublished internal Science Museum report produced by Creative Research Ltd. (1998).
15. Comments from the discussion exhibits are reproduced exactly as they appeared with regards to spelling, capitals, underlining, etc.
16. *Future Foods?—Visitor Comments*, an unpublished internal Science Museum report, ed. Nicola Perrin (1998).
17. *Join the Great Fat Debate*, an unpublished internal Science Museum report, ed. Xerxes Mazda (1998).
18. Douglas Worts, "Gallery enhancement: Forging a new partnership with the public," in *Visitor Studies: Theory, Research and Practice 6* (Jacksonville, Alabama: Visitor Studies Association, 1993), pp. 176–197 (p. 180).
19. *Big Bang—Visitor Comments*, an unpublished internal Science Museum report, ed. Sarah Hunt (1996).
20. *Future Foods?—Visitor Comments*, an unpublished internal Science Museum report, ed. Nicola Perrin (1998).
21. Ibid.
22. *Join the Great Fat Debate*, an unpublished internal Science Museum report, ed. Xerxes Mazda (1998).
23. *Evaluation of Future Foods?*, an unpublished internal Science Museum report produced by Creative Research Ltd. (1998).
24. *Future Foods?—Visitor Comments*, an unpublished internal Science Museum report, ed. Nicola Perrin (1998).
25. Yvonne Harris and Ben Gammon, *Join the Great Fat Debate summative evaluation*, an unpublished internal Science Museum report (1998).
26. *Join the Great Fat Debate*, an unpublished internal Science Museum report, ed. Xerxes Mazda (1998).
27. *Future Foods?—Visitor Comments*, an unpublished internal Science Museum report, ed. Nicola Perrin (1998).
28. This view was not unusual—see Les Levidow, "Domesticating Biotechnology: How the London's Science Museum has framed controversy," *European Association for the Study of Science and Technology Review*, 17, no. 1 (1998), pp. 3–6.
29. *Future Foods?—Visitor Comments*, an unpublished internal Science Museum report, ed. Nicola Perrin (1998).
30. *Evaluation of Future Foods?*, an unpublished internal Science Museum report produced by Creative Research Ltd. (1998).
31. *Future Foods?—Visitor Comments*, an unpublished internal Science Museum report, ed. Nicola Perrin (1998).
32. *Bucky Balls* aimed to introduce visitors to the science and potential applications of Bucky Balls, a recently discovered form of carbon.

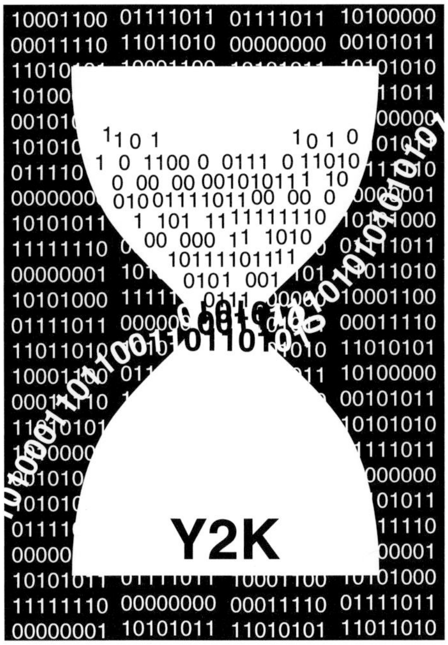

"A-OK" to Y2K

TECHNOLOGICAL CONFIDENCE AND HISTORY AT THE END OF THE TWENTIETH CENTURY

Robert Friedel

The discussion that follows is not, I want to admit at the outset, primarily about museums or the public understanding of science—at least on the surface. It is about history and how it is understood and used. It is, more specifically, about relations between historical understanding and a widely perceived and discussed technological problem, how computers will be affected by the century's turn. Nonetheless, I contend that if our museums of science and technology are to be honest and effective vehicles for examining and teaching about the world that their visitors confront, then new attitudes toward history and its place in the science museum are needed.

The role of history in museums of science and technology is both explicit and implicit. It is explicit in exhibitions that seek to explain "where things came from." Perhaps more significantly, history is implicit in exhibitions that seek to explain both "how things are" and "where things are going." This latter, implicit, role is often difficult to see, but it is in fact fundamental to the way that we depict the modern world. I want to make this point, not by examining museums or exhibits themselves, but by looking at one of the most talked-about technological circumstances of our time. In doing so, I want to tease out lessons about the relationship between modern technologies and their histories. But my claims are not simply that modern situations are the product of historical experience—this is rather trivial. Instead, I want to make the case that today's technologies are in some measure the product of the way we *perceive* this historical experience. This has profound implications for the museum's role in shaping that historical perception.

The natural process of reflecting on our passing century involves much pride in technological achievement—it is, after all, the century of wireless communications, jet travel, space exploration, and computer networks. Marvelous, therefore, is the irony that surrounds all this in the form of the "Y2K Problem," or the "Millennium Bug." This century of technological triumph, in which the creation of imposing instruments of data manipulation and exchange has been perhaps the most surprising, least expected, of all our technologies, ends with near panic at the prospect of malfunctioning machines and systems all over the globe, ranging from the elevator down the hall to the world-wide air control system. In mid-1999 it is too early to tell just how far this panic is unjustified hysteria and how much of it is legitimate concern, but already we know that billions of dollars—and euros, pounds, yen, and the like—have been and are being spent to control the problem. Alongside this is the accompanying flood of warnings, pronouncements, prescriptions, and analyses: entire shelves of books, reams of articles, and more Web sites than can be counted have been devoted to Y2K. In the midst of this flood, it's a bit surprising how little attention has been paid to the historical significance of the phenomenon, both in the sense of its longer-term importance but also in the sense of what it can tell us about history and our historical sense of the technologies around us.

The Y2K problem, in fact, only makes sense in the context of the history of a particular technology—that of the digital computer—and in light of how changes in that technology and all modern technology are perceived and characterized. It is an extraordinary opportunity for students of technology and its history, and for those who would exhibit that history in the museum or explain it to a wider public, to explore truly fundamental questions about technological change and technological confidence. I will argue, in fact, that this opportunity gives us a chance to educate ourselves and the public about not only a technology of surpassing importance in modern life but also about the most fundamental relationships between past and present.

My interest in what we historians might have to say about Y2K was sparked by reading journalists' accounts of the situation. As an example, the following comments appeared in a guest editorial column in the Washington *Post* a few months ago:

> The year 2000 (or Y2K) Problem is a glitch that makes computers confused about the current date. It has also caused worries that are now blossoming into full-blown, completely unwarranted hysteria.
>
> . . . Why Y2K? In a frugal but misguided attempt to save space, many computer programs were written to express dates with only two digits; thus, '98' for 1998.[1]

Subsequent observation confirms that this kind of explanation is a common one. This particular author goes on to insist that paranoia and hysteria are emerging from all this, and to urge cooler heads. But this got me to thinking about his explanation—"a frugal but misguided attempt" by computer programmers? When I asked a historian colleague whose first career was indeed as a computer programmer in the late 1960s and early '70s, he agreed heartily with the "frugal" part, but most certainly not with the "misguided." I'm certainly willing to go along with that, except that it raises another, even more interesting, set of questions: Is the Y2K Problem the result of one or more "mistakes"—steps taken or decisions made that could, with appropriate knowledge or foresight, have been avoided? If so—what mistakes, by whom, and when? If not—does this mean that this is a completely unavoidable situation, an inevitable product of computer technology? Or does the answer lie in some kind of middle territory—not resting on the mistakes of particular decision-makers or technicians, nor somehow fundamentally part of the technology of the information age, but instead based on the values and expectations of modern technological civilization? I think that if we can answer these questions, we've shown what a constructive crisis this really is for our understanding of technology.

Further observation of the Y2K literature shows a number of characteristic accounts for the origins of the problem, although some divergence about blame or accountability. It is widely understood that saving on computer memory usage in the early days of computing is at the root of the situation, but from this point there is an interesting range of opinions. The suggestions include:

- Programmers were simply stupid.
- Programmers (or those who paid them) were greedy and short-sighted.
- Programmers were infiltrated by the KGB (really—this in an American tabloid), who plotted to disrupt the West's computers at the end of the millennium—or some similar paranoid image.
- Programmers knew this would be a great money-making opportunity, as corporations and governments would be desperate for their services.

The answer does not lie with these notions of stupidity, duplicity, or conspiracy, but rather in the nature of technologies and of the decisions that shape their experiences and their places in the economy and society at large.

To understand how we can make sense of this, we must first examine the basic technical elements of the key programming decisions involved. In addition we have to recognize some elemental features of technologies and technological systems, particularly in the early stages of their development. One basic principle was pointed out recently by Smithsonian computer historian Paul Ceruzzi, who began his own explanation of the Y2K situation by noting that technologies frequently just barely work when they are being developed. They are not designed to some sort of theoretical optimum, but rather inch their way toward useful functionality by small steps that are often crude and tentative. The first electronic digital computers, devised in the mid- and late 1940s, depended on vacuum tubes, not transistors, and they broke down after only about 15 or 20 minutes of operation due to a tube burning out. A fundamental measure of technical progress of early complex electronic devices was the term "*mtbf*," which stood for "mean time between failures." Even through the 1950s, the *mtbf* of a computer was often measured in days or weeks. These were still "tender" machines, requiring enormous patience and much care. Just getting them to work at all for any length of time was the operators' and the programmers' key aim. They also had very little memory, and input and output devices were slow (by today's standards) and also prone to failure.[2]

This fact has been much neglected in Y2K discussions of early computing, and yet it is fundamental. It is also fundamental to the history of many complex technologies, and yet its implications are ignored. If getting a machine or system to work is very difficult, then the primary value in design lies in its working at all, not in its achieving some ideal. Once a working design has been achieved and proves itself robust, then features of that design will prove difficult to dislodge. In fact, these features will typically be changed only when they must, and not before.

In the case of early computers, additional constraints pushed design of both hardware and software in particular important directions. In the first computers, program and data storage was physical and mechanical, embodied in paper tape or punched cards. The 80-column Hollerith card (or "IBM card") was, in fact, the symbol to an entire generation of youth of computer regimentation and its limitations, with the adage "Do Not Bend, Fold, Spindle, or Mutilate" representing the demands and constraints that the computer placed on its users and subjects. For our purposes, this also neatly represents the technical limitations of these original storage systems. It is small wonder, therefore, that every incentive existed, from the beginning, to reduce data entry and size to a minimum.

Even when memory technologies changed to magnetic storage, costs were high and there was every incentive to keep programmers' codes short. Magnetic disk storage, introduced in the early 1960s, was a wonderful boon to computer technology, and its utility helped to increase computer usage (and reliability) enormously. But in 1963 it still cost about $175 per month to lease 1 MB of disk storage on a computer—that works out to more than $10,000 per year in current (1995) dollars. Compare this to our costs of less than 10 cents per month per leased MB on a modern mainframe computer—the difference is 10,000 times. By the early 1970s, storage costs had fallen to about $1,600 per MB per year in current dollars—still 1,500 times today's rate. Ten years later, in the early '80s, the cost had dropped more substantially—to about 22 times current costs.[3]

Some rather crude, but still plausible, calculations are needed to gauge just what kind of difference was made by accommodation of shorter dates. In the American style, a complete date is written: 05/28/1999. Early programmers quickly dropped the separators (the /'s) and truncated the year, yielding: 052899. This is an elimination of 40% of the original number of characters. The truncation of the year alone is half of this savings. It has been estimated that date fields may occupy as much as 3% of all computer data (at least in early years, when "data" meant essential information, such as names, addresses, and a variety of dates, rather than extensive text fields). The elimination of the century figures ("19") from dates could thus have yielded savings in data space on the order of 1%. In light of the high memory costs cited, this is not at all a trivial savings, particularly over a period of years or even decades. The decision to truncate dates in computer code was unquestionably a rational strategy for saving scarce resources; to have done otherwise would have been wasteful.[4]

Over time, from the earliest computers and programs, memory became cheaper and the remaining years of the 20th century became fewer. The turning of the digits at this year's end is not exactly an unpredictable event, thus the real historical question—ignored by almost all discussions of Y2K—is not why early code was written with two-digit years, but why it persisted. This persistence, more than the original decision to save memory space on dates, is seen as the "mistake" behind Y2K, and, to be sure, there is clearly some point at which it *is* a mistake to write computer programs that will fail in a number of years. But that point is indeterminate, and the long life of computer code and of programming practices that fail to accommodate Y2K points toward a number of important lessons:

- Short-term returns will almost always win out over long-term solutions, particularly in competitive environments.
- The persistence of computer code—short date fields and all—is not an accident or error; it reflects a nearly universal if not in fact necessary aspect of the way all technologies develop.
- While the century gets shorter and 2000 comes closer, there are other forces at work that make the old code *more* valuable rather than less.
- The economy of using old code and old standards for dates persists even when memory costs are no long very relevant.

Understanding these less-than-obvious aspects of the story brings us much closer to an appreciation of the historical lessons from Y2K.

I am not myself a computer historian, and I have little or nothing to add to computer history *per se* in this discussion. It is worth pointing out, however, that there are some specific elements of computer history that are relevant to understanding the Y2K situation. Paul Ceruzzi recently pointed out one of these: One of the most important commercial breakthroughs in computing history was the introduction in late 1963 of IBM's System/360 computers. This was a significant redesign of the mainframe computer, taking full advantage of transistorized circuits and disk memory storage. IBM designed a range of 360's to fill market demands from large corporations to small companies, universities, and government agencies. So expanded were the capacities of the new machine that it would not have been out of the question to accommodate four-digit years in the basic code, and thus avoid any Year 2000 problem. But at this point IBM was in heated competition for the business market with the Honeywell Corp., and when it looked as though the System/360 was going to be introduced in such a manner as to require significant amounts of new programming for its users, Honeywell announced an inexpensive but powerful new machine designed to emulate the older basic IBM model, the 1401. This emulation called for 2-digit year coding and all the other memory-saving tricks of the older computers.

The Honeywell H-200 began selling well, and IBM became fearful of a significant loss of sales. They responded by developing versions of the System/360 that emulated the 1401, allowing customers to use old programs, while at the same time exploring possibilities of the new architecture. This worked brilliantly both from an engineering and from a commercial point of view, and IBM's dominance of the mainframe computer market was secured. But the emulation of the 1401—what we now call "downward compatibility"—locked in the two-digit year codes for another generation. One important lesson here is that short-term advantages, in the real world of business and engineering, almost always triumph over long-term ideals. As we shall see in a second, this should not surprise us one bit, but we neglect this fact all the time in our interpretations of modern technology.

Were computer programmers aware of what they were doing? Did they believe that the programs they wrote in the 1960s were not relevant to what would happen thirty or forty years hence? They certainly were aware that computer code is very valuable and very persistent stuff; programmers learn early to recycle old code whenever they have the chance. Not only does this save initial effort, but it also saves debugging time further down the road, since the more tried and true code you use, the less the chance of new errors in a program. Just getting a computer's programs to work at all is typically the programmer's core mission; anyone who has memories of debugging even a simple BASIC program will appreciate this fact. This leaves very little room for the niceties of worrying about the implications of recycled code or what might happen a few years down the road.

As paleontologist and historian Stephen Jay Gould has pointed out, this is exactly how nature works in the course of evolution. New species always come from old ones, and in the course of evolution useful patterns and forms persist from species to species. This persistence is economical, although it often carries with it un-useful consequences. The nipples on a male's chest, for example, have no utility whatsoever—it just happens to be useful to nature to operate with a single human plan which is, in the course of sexual development, differentiated into a useful female variety—the useless male nipples are simply artifacts of that undifferentiated plan.[5] The appendix in humans has no useful role; it too is a historical marker—a bit like old computer code. It's often simpler (and more likely) in the course of biological and technological development to use tried and true plans, even if not every aspect of the old plans is relevant to the new.

There are other ways in which the past lives in the present, even in the rational, planned world of technology, and appreciating these is an important step to liberating us from the trap that led to Y2K. One of the most notable technological examples, used by Gould himself, is the typewriter and computer keyboard—in English-speaking countries, the QWERTY keyboard. Stanford economic historian Paul A. David gets credit for being the first to look systematically at the history of this apparently sub-optimal technology and draw lessons from it.[6] The keyboard is "sub-optimal" in the sense that it is actually designed to slow down, rather than speed up, typing. This was due to the nature of the original mechanical typewriter, as designed by Christopher Sholes in the 1870s. Sholes discovered that if letters physically close to one another in his typewriter were struck consecutively, they were likely to jam, that is, the typebars would get in each other's way as one came down from striking the platen and the next came up. The sensible design solution was to determine what letter pairings in English were most common, and to separate those pairings in the keyboard layout. The resulting QWERTY keyboard, as introduced by the Remington typewriter makers, actually speeded up overall typing speed, but by reducing mechanical jamming rather than by enhancing the typist's ability to hit keys in a rapid rhythm. As typewriter mechanisms and construction improved, first in mechanical devices and later in motorized and then electronic typewriters and word processors, the virtues of the QWERTY layout diminished and then disappeared. Alternative designs have been introduced over the past century, but to little avail.

Professor David refers to the phenomenon by which historical circumstances, now no longer operable, determine current conditions as "path dependence." This means simply, as he puts it, "where things are heading depends upon where they have been." Gould, for his part, breaks the phenomenon down into two distinct elements, which he calls "contingency" and "incumbency." Contingency refers to the fact that the path of development or evolution is often determined by specific circumstances, not in any particular way inherent in the system but the product of chance or, at least, of a confluence of conditions quite unpredictable from the initial state of the system. Incumbency refers simply to the fact that things tend to remain as they are and have been unless some force or circumstance works to effect change. In David's language, some systems are governed or influenced by "positive feedback," so that a given tendency tends to reinforce that tendency over time. This is particularly important language for economists, for economic systems tend to be described largely in terms of "negative feedback," systems that are "self-correcting" in some way, or achieve equilibrium by competing and countervailing forces (supply and demand being the most obvious example). One result of focusing on negative feedback, David points out, is to construct ahistorical explanations for things—in dynamic systems that are constantly self-adjusting, the past states of the system are not very relevant. Positive feedback, on the other hand, makes history very relevant indeed.

The Y2K problem is clearly an example of incumbency at work: the two-digit year code was a good solution to a very real technical problem (just as the QWERTY keyboard was a good solution to the problem of jamming typewriter keys). It outlived its usefulness once computer memory became cheap (just as the keyboard did with the coming of electric typewriters, not to mention computers). But it was difficult to wean ourselves away from the convenient shorthand of the two-digit year, until near panic began to move things. David himself has pointed out that just as much to blame for the Y2K crisis as initial programming decisions is the modern engineering mindset known by the acronym "JIT," for "just in time." This attempts to maximize productivity by putting off actions as long as possible. Once resources have been allocated to a particular expenditure or set of actions, they cannot be recovered, goes the logic. If circumstances change or better solutions emerge, then resources have been squandered or, at least, spent less efficiently than they might have. So JIT engineering tells us to wait until alternative actions are no longer possible. There is a degree to which this often happens anyway, even without conscious engineering decisions, for a competitive system will almost always favor relatively certain short-term returns over more problematic long-term ones.

This tendency to think only in the short term is encouraged by another key element in the Y2K story: the belief that technological change will sweep away old problems. The twentieth century has seen its share of shocks to technological optimism, but these have proven generally limited and short term. For the first two-thirds of the century, at least, technological confidence was very much the order of the day in the West. This contrasts with sources of pessimism, such as World Wars and Depressions, that make the twentieth century version of optimism rather different from its Victorian predecessor. Indeed, it's important to understand that while the assumptions of inevitable economic and social progress that accompanied Victorian technological swagger were dealt mortal blows by the shocks of our own century, confidence in technology itself has rarely been challenged until our own generation. The apogee of this confidence can be conveniently marked—at least for Americans—by Apollo 11's landing on the surface of the moon, just short of 30 years ago (July 20, 1969). Astronauts assuring us all that everything was "A-OK" was the perfect image of technology in and under control. Until the Challenger disaster of January 28, 1986, the technologies of the Space Age were the consummate symbols of confidence (even in the face of occasional setbacks) and they extended themselves more widely with the oft-repeated sentiment: "If we can land a man on the moon, why can't we . . . ?" Fill in this blank with a whole range of worthy goals, technological and otherwise.

It was in the midst of this exuberant confidence, it can be argued, that the key Y2K decisions were being made. It seems a little startling to think of it now, but in 1969 computers were still not a very large part of the scheme of things, at least not by today's measure. They were certainly still in their adolescence if not infancy, and the logic of history to the computer makers of the day was clear: computers and programs could be expected to continue to change radically for the foreseeable future. In some ways their anticipation was correct. As we've seen, the cost of computer memory continued to decline dramatically, and this, along with the declining cost of other electronic elements of the computer, did indeed transform the computer physically and functionally over the next decades. But—and this is the interesting point—it did *not* make obsolete all key elements of the system of computer technology. Indeed, the increasing importance of computers, due in large part to their declining cost, made certain elements of the system ever more valuable—particularly reliable computer code. In particular, software that worked reliably and could be used flexibly came to have ever greater value. Even as the year 2000 approached, in other words, good old code became more, not less, valuable. Why else is Microsoft, not IBM, the corporate gorilla of the computer age?

One of the interesting questions this raises is the relationship between technological improvement, on the one hand, and obsolescence, on the other. The images and metaphors that we've been asked to adopt for change in complex systems—say that of the "advancing front" with its occasional salients and reverse salients—assumes that advancement of the front will inevitably entail change all along the system. This is not, in fact, always—or maybe even generally—the case. It is not at all unusual for systems to accommodate improvement in some areas while not in others. The "unimproved" areas are not necessarily "reverse salients," awaiting catch-up by inventors and developers, but often are adequate, useful, economical elements of the system that resist obsolescence. But this is not, in fact, part of the models of change we carry around with us. Indeed, it's possible that over-attention to the dynamics of systems has misled us into thinking that localized improvements must eventually yield change all through the system. Even when improvements are important, this simply is not necessarily so. Incumbency is a fundamental fact of technological life and change.

We have perhaps grossly understated the rational, economical, sensible conservatism of our technological world. This is particularly true, I would argue, in our characterization of the technologies of the 20th century. There is a widespread belief, almost desperately clung to in some quarters, that technological change is necessarily an accelerating phenomenon, and that change in our own time is thus faster and more thorough than at any time before us. This is a fundamental element of the technological confidence I've referred to—if change is ever faster, then the likelihood that "problems" will fall before the assault of our scientists and engineers is naturally always increasing.

This idea of accelerating technological change is older than the 20th century itself. The anthropologist Lewis Morgan wrote in 1877 that "Human progress, from first to last, has been in a ratio not rigorously but essentially geometrical (slowest in the first period, and most rapid in the last)." Almost 50 years later the sociologist William F. Ogburn came to a similar conclusion that "From neolithic times to the historic period and from the historic period on, the changes in material culture have been much more rapid. At the present time both the change and the accumulation of material culture are quite rapid and may be measured in such brief intervals as generations or even decades." Ogburn was writing in 1922; his pupil, Hornell Hart wrote in the mid-1950s that "man's power to carry out his purposes in the material world has been increasing faster and faster. The reality of acceleration in technological growth can scarcely be doubted by any intelligent person who is acquainted with the facts."[7] Hart supported this claim with graphs and charts about changes in everything from land speeds to war casualties, and then went on to account for acceleration largely through the fact that "every invention consists of a new combination of old elements." The logic here is simply that invention is, by its nature, a kind of multiplier effect, and therefore must increase geometrically.

These sorts of claims are all around us. In a piece in London's *Daily Telegraph* last July (London, July 16, 1998, p. 10), Peter Cochrane, a British Telecom research manager, proclaimed, "We are surfing [!!] a wave of change that is getting bigger and faster than ever before. . . ." Earlier this year, a section of *The Economist* devoted to "Innovation in Industry" (Feb. 20, 1999, pp. 7–8) discussed the dynamics of innovation in Schumpeterian waves of "successive industrial revolutions" with the conclusion that the cycles are shortening—from Joseph Schumpeter's 50–60 years to something closer to 30 years today. Indeed, the notion that change is ever swifter and more profound as we head toward the future is so commonplace as to hardly require comment.

The danger of this idea is made wonderfully apparent by the Y2K problem. To be sure, the first makers and users of digital computers in the 1950s and 1960s did indeed witness breath-taking change, and certainly machines and practices in 1970, say, bore little resemblance indeed to those of 1950. The computer of 1990 was also rather different from that of 1970, but the differences were more a matter of scale than of kind, and while scale (memory, speed, interconnections, etc.) counts for a great deal in computing, change itself is not of the same character as in the early years of the technology. This, too, is very common in the life of the devices and systems around us, but its significance is much neglected. The Y2K problem brings home powerfully the real implications of differing qualities of technological change. The sizable literature on the problem uses the term "legacy systems" to refer to computers and programs that are not Y2K compliant due to their retention of old design practices and assumptions, as well as old programs and machines. The historian has to smile a bit at this term; "legacy" simply means that history is still with us.

The importance of the kind of mentality that attempts to relegate "legacies" to history's dustbin can be illustrated by a little more detail on the Y2K problem's own history. The most important programming language for many of the systems causing concern is COBOL, a business-oriented language devised at the beginning of the 1960s, largely under U.S. government sponsorship. The government wanted a programming language that was not owned and used exclusively by one manufacturer, but which could be adapted for many different computers. The core of the COBOL (**Co**mmon **B**usiness **O**riented **L**anguage) that resulted was a compiler that Grace Murray Hopper wrote for the UNIVAC computer. Other companies' business-oriented systems were integrated into this, and the government quickly pushed the result—written by a multi-company committee—into a standard, requiring its use in government-purchased machines. One of COBOL's great distinctions was that it made possible the use of ordinary English expressions in instructions and data definitions. This gave it a "natural" look and feel (at least as compared with the heavily symbolic languages commonly used) and made it possible to be "self-documenting" (that is, it didn't require so many explanations written by the programmer as comments to explain what lines of code were supposed to do).[8]

The real significance of COBOL for the Y2K story lies in its success and the legacy that it left. Used in the computers of all major companies, mandated in its machines by the U.S. government, relatively easy to learn and write, COBOL systems became, in Gould's term, "incumbent," and even through the great technical changes of the next several decades of the computer age, COBOL programs and their derivatives were everywhere. But COBOL was written in the days of punched cards and of 10 cents-a-bit memories. It could, particularly thanks to something called the "PICTURE" clause, handle date fields of any length, but programmers only took advantage of this when they had to (elders of the Mormon Church, for example, needed it for genealogical records). Thousands, and then millions, of lines of COBOL code were written with two-digit years. A few of the COBOL designers pointed out the limitations of this early, but standardization, not adaptation, was the spirit behind its use. The Pentagon mandated the two-digit year in all code, and the standard—once a compelling economy, but later an anachronism—became nearly universal. This is how legacies are created in the computer world.[9]

Legacies are not irrelevant to understanding today's and even tomorrow's technological world. Path dependence is, in fact, everywhere, but we do not say this very clearly in the histories we write or in the exhibitions we display. This is pretty peculiar, from one perspective, since the relevance of history would seem to be central to the mission of historians and historical museums. The picture of the scientific and technological past that we project, however, tends to be of a dead past, of ideas and machines that have been superseded by their betters. The past is viewed as simply steppingstones to the present rather than integral parts of it. Even when we introduce or include questions about the suitability and desirability of our present condition, the image of the superseded past, whether viewed as better or worse, remains.

Imagine, if you will, a different attitude in our histories and displays. Here questions would be asked both about what has changed and what has not in our technologies. The existence of alternative paths—the gas refrigerator, the electric car, the earth toilet—would make clearer the nature of choices made in the past that still guide our present. Historically taking apart devices and systems would reveal the necessary intermingling of old and new as they change over time. Such parsing would no doubt make clearer the extent to which our technologies always tend to be compromises rather than ideal systems. The sheer difficulty of simply getting something novel to work for the first time, and then to work in a manner the market will tolerate and accept, could be appreciated, with the result being a better understanding of what our technologies don't do as well as what they accomplish. Thanks to the work of the "social constructivists" and the "contextual historians" of recent years, we understand much better than we once did just how contingent many devices and systems are, how they are the products of competing agendas and values as well as the realization of physical possibilities. We need now to put that understanding to work. A museum that confronted these issues squarely and honestly would be one that could make a lasting and wide-ranging difference.

The past is, in the phrase adapted by David Lowenthal, a foreign country, a territory made distinguishable from the present by not only time but difference. We are right to probe into and attempt to explain the differences. But, in doing so, we need to be more careful about linking past with present, and marking the persistence of things as well as their passing. The made world is not the product of plan or logic, though they have their place. It is much more the product of historical process. Our technologies contain within themselves their histories, and they are unimaginable and unfathomable without an understanding of them. At century's end, our real challenge may be to muster the candor and honesty needed to confront this simple fact.

NOTES

1. James K. Glassman, "Bonkers Over Y2K," *Washington Post*, Dec. 1, 1998, p. A25.
2. Paul E. Ceruzzi, "Y2K: Old Hat in New Technology," *Washington Post*, Jan. 11, 1999, p. A19.
3. Leon Kappelman and Phil Scott, "Accrued Savings of the Year 2000 Computer Date Problem," at http://www.year2000.unt.edu/kappelma/accured.htm (Oct. 10, 1996). Another historical measure of memory costs can be found in E. W. Pugh et al., *IBM's 360 and Early 370 Systems* (Cambridge, Mass.: MIT Press, 1981), p. 475; thanks to Emerson Pugh for pointing this out to me:

   Manufacturing costs per bit of main memory went from \$.70 in 1953 (CRT), to \$.08 in 1959 (Core), to \$.012 in 1971 (Bipolar IC) to \$.0002 in 1979 (SAMOS IC). This is a decline of a factor of 3,500 even before the plunge in chip production costs of the 1980s. From 1979 to 1997, the cost per bit for one of the most important types of memory chips, the DRAM, fell an average of 68% *per year* (this is about a factor of 500 over the 18 years). *Status 1997*, ed. Bill McClean (Scottsdale, AZ: Integrated Circuit Engineering Corp., 1997), pp. 7–62; my thanks to Nance Briscoe of the Smithsonian for this source.
4. This point is made most clearly in Leon A. Kappelman and James J. Cappel, "Critical Issue Report: Facing the Year 2000 Problem." My thanks to Prof. Kappelman for sharing this with me. It can be found at http://www.year2000.unt.edu/kappelma/critical.htm. Kappelman and Cappel go on to estimate the savings actually represented by programmers' choices:

   "It is estimated that conserving precious disk space with the two-digit year format saved the typical organization over \$1,000,000 per gigabyte of total storage in the 30 year period from 1963–1992 (in 1992 dollars). Factor in the benefits (and presumably profits) generated by these unspent dollars, using a conservative 10% internal rate of return, and the opportunity cost per gigabyte of total storage rises to more than \$15 million. These savings estimates can be doubled for date-intensive areas like banking, insurance, securities, credit, and others."
5. Stephen Jay Gould, "Male Nipples and Clitoral Ripples," in *Bully for Brontosaurus* (New York: W.W. Norton & Co., 1992), pp. 124–38.
6. David's original article is "Clio and the Economics of QWERTY," *American Economic Review*, 75 (May 1985); this was expanded in "Understanding the Economics of QWERTY," in *Economic History and the Modern Economist*, ed. W.N. Parker (London: Blackwells, 1986); recent contributions on path dependence include "Path Dependence and the Quest for Historical Economics: One More Chorus of the Ballad of QWERTY," Discussion Papers in Economic and Social History, Univ. of Oxford, no. 20, Nov. 1997. On the relationship to Y2K, David has contributed some remarks to the Economic History—Research Listserv, EH-RES: "Qwerty and the Economics of the Millennium Bug" (17 Dec. 1998), archived at http://www.eh-net/ehnet/Archives/eh.res/dec-98/0019.html.
7. All quoted in Hornell Hart, "Acceleration in Social Change," in F. R. Allen et al., *Technology and Social Change* (1957).
8. Marjorie Berk, *The Programmer's COBOL: A Complete Reference* (New York: Interact Publ., 1970), particularly the foreword by Robert Bemer, pp. vii–xv; Jean E. Sammet, *Programming Languages: History and Fundamentals* (Englewood Cliffs, NJ: Prentice-Hall, 1969), pp. 330–377; Paul E. Ceruzzi, *A History of Modern Computing* (Cambridge, MA: MIT Press, 1998), pp. 91–94.
9. Robert Sam Anson, "The Y2K Nightmare," *Vanity Fair* (January 1999); Chris Taylor, "The History and the Hype," *Time* (Jan. 18, 1999): 72.

Communication and Communities

CHANGING PARADIGMS IN MUSEUM PEDAGOGY

Eilean Hooper-Greenhill

Museums in Britain are experiencing a revival of confidence and a flowering of new opportunities. In part, the museum renaissance is fueled by the National Lottery. New museums, or museum-related institutions, seem to be opening almost daily: the National Centre for Popular Music in Sheffield opened in January 1999, and the Earth Centre in Doncaster opened in April 1999. Established museums are opening or planning new wings or new branches—one example is the Neptune Court at the National Maritime Museum, which opened at the end of March 1999 and is intended to redefine what a maritime museum might be. Some of the upcoming projects have similar ambitions. The Spiral project at the V&A, for example, designed by Daniel Libeskind, will celebrate creativity and inspiration rather than display objects. The New Art Gallery at Walsall will enshrine a new and more proactive relationship with audiences within the architecture itself. However, the rate of development and the number of new ventures poses its own problems over whether there will be sufficient visitors to go round. Early indications suggest that unless the experience offered meets the perceived needs of visitors, targets will not be met. New opportunities must be matched by new visions and new professional practices.

The second driver of the museum renaissance is government policy. Since coming to power, the Labour government has developed detailed cultural policies that stress firmly that continuing public subsidy will depend on demonstrable evidence that museums and the arts are for the many, not the few.[1] Chris Smith, the Minister for Culture, has said, "Financial allocations will be closely tied to outcomes which reflect our four central themes—access, excellence and innovation, education and the creative industries."[2] How these outcomes are to be demonstrated is the subject of some debate, but the emphasis is squarely on "the public rather than the producer." The Department for Culture, Media and Sport intends to be proactive to ensure that their policy is delivered by those museums that they fund. To quote Chris Smith again: "The Government has provided the resources and the will, but we know we cannot just sit back and hope that these are transformed into better and more accessible performances, sporting records, improved cultural education and more opportunities for the excluded. We will give direction; we will set targets and chase progress; and where appropriate we will take direct action to make sure that our objectives are achieved."[3]

In many ways, DCMS is pushing at an already open door, but these kinds of statements, and they are heard frequently, do focus the mind. However, current government priorities build on the shifts in museum culture that have been in evidence over the last 10 to 15 years. During this period, museums have developed more sophisticated approaches to communication, have begun to analyze audiences more carefully, and have accepted the necessity for differentiated provision. The management of museums has improved, and the professionalism of museum employees has become more developed. Documents such as Victor Middleton's two reports,[4] with their discussion of the economic challenges and possibilities facing Independent museums, have had an enormous influence in Britain. Many museums in the UK have, over the last few years, demonstrated entrepreneurial flair, imaginative thinking and managerial acuity.

"Museum" is a capacious concept. Twenty years ago it seemed that there was only one way to conceptualize what a museum might be. The worst of these were elitist, uncomfortable and frequently authoritarian; the best were "worthy but dull." Where innovation occurred, it was not always welcomed or acknowledged at a policy-making level. There are still museums that might be open to charges of this nature, but there are many more that are finding new ways to operate. The concept of what a "museum" might be is enlarging.[5] The museum sector in the UK now is very diverse, high profile, and energetic, with more open, exploratory and imaginative ways of working. The emphasis on scholarly knowledge and the value of researching and learning from collections is retained, but these older inwardly focused values are complemented by more outwardly oriented values that are leading to creative partnerships with audiences, sponsors, and other cultural agencies and providers.

However, these changes in museum culture have not gone unchallenged. Some have seen the more open approach to audiences as regrettable, and a more entrepreneurial managerial style as inappropriate. Museums have been charged with being too much like "theme parks," with "Disneyfication," and with "dumbing down." How can these charges be refuted? How can they be analyzed? What do they really mean? A more developed analysis of the museum-audience relationship will enable a better understanding of past and current museum practice. Cultural and communication theory is useful in trying to develop such an understanding. This paper discusses two paradigmatic formulations of the museum-audience relationship. The transmission approach to museum pedagogy is seen as characteristic of an older model of what a museum might be, a model that developed during the modern period, but which is now charged with having served its purpose.[6] The cultural approach takes a broad and egalitarian view of the pedagogic role of museums which focuses on the meaning-making activities of differentiated audiences and which forms the basis for the development of new professional practices.

MODERNIST MUSEUMS AND THE TRANSMISSION OF KNOWLEDGE

The nineteenth-century public museum has been called one of the emblematic institutions of the modern period.[7] The period that we now call the "modern period" is frequently difficult to pin down in terms of specific dates. Its beginning can vary from the seventeenth century to the beginning of the nineteenth century, depending on whom you read. Its end is more commonly agreed to be in the post-war period—around the 1950s–'60s. There is, however, no agreement as to whether we are in a period of late modernity, or post-modernity. In some ways, the characterization of modernity as an "unfinished project" is attractive. There is no doubt that many of the aims of the modern period have not been achieved, and were probably never achievable. The modern age was the age of capitalist development, industrialization, and colonialization.[8] Its major epistemological achievement was the development of positivism.[9] Many of the aims of the modern period have been severely called to task, but equally it is clear that many of the basic assumptions of the modern age still influence us deeply today.

The modernist museum was European in origin, although this model would be exported worldwide. Authoritative, the museum was tasked with the production and dissemination of knowledge that was constructed through a number of collection-related disciplines (art, natural history, geology, archaeology, ethnography). Through the relationships between people, words and things, the modern museum was also charged with converting raw humanity to civil society.[10] For example, in 1805, Martin Archer Shee, later to become the President of the Royal Academy, referred to the arts, including museums, as "those softeners of human life, those refiners of the rough, drossy ore of humanity . . . ".[11]

In the rude industrial world, the arts represented something finer, more elevated, something to look up to and to aspire to. Museums could offer opportunities for self-education and self-elevation at a time when schooling was not compulsory for all and when there were few other opportunities for self-improvement. However, the model of teaching and learning that museums were implicitly using took little account of the individual needs of learners. It was assumed that if collections were laid out in such a way that relationships between them were made visible, the underlying knowledge structures would be clear to the viewer and therefore learning would happen.

Thus, in the galleries of the British Museum (Natural History) during the 1880s the principles of the classification of birds and beasts automatically became the principles of display design.[12] Similarly, in art museums, the works of art presented the histories of art.[13] The relationships between the items in the collections, rather than the objects themselves, formed the subject matter of the lessons of the museum. A walk through the museum galleries meant a walk through the structures of specific disciplines. It was assumed that, having carried out the walk, visitors would understand the relationships between the objects, and that this relationship, carefully crafted in the display itself, would be both visible and comprehensible to the viewer/visitor.

The public museum in the nineteenth century was intended to appeal to a new mass audience. Its role in developing citizens was educational. The approach it took to communicating with its audiences, and to educating its visitors, was through the visual display of collections. Later this was to be supplemented with information that would enable a better appreciation of the relationships that the visual arrangements of the collections embodied. The display, with its informative texts, became akin to the academic lecture, but with the lecturer absent.[14] Museum displays were used as a method of teaching at a distance a mass audience that through education would become both more individualized and more susceptible to self and social discipline.[15]

MASS COMMUNICATION AND BEHAVIORIST PEDAGOGY

Mass communication has been the subject of considerable analysis. It is described as being based on what has come to be known as a transmission model of communication.[16] The process of communication is understood as a message that is transferred from a communicator through a medium to a receiver. This approach to communication is based ultimately on a behaviorist view of education.[17] Models of teaching and learning and models of communication are inter-dependent. Both encapsulate views about the nature of knowledge, the processes of learning and teaching, and the nature of the learner. In the behaviorist approach, education is understood as a process of stimulus and response, where the teacher provides information through a stimulus, which the learners absorb and to which they respond.[18] The teacher is the source of knowledge and authority within the pedagogic process, and the task of the student is to learn what is taught. The student is assumed to bring nothing to this process, and is described as a "tabula rasa" or "empty vessel."

In the museum, the curator, the scholarly authority, transmits information about science, art, or history through the medium of collections presented in the form of a display to the receivers, the museum visitors. In some ways this describes in an apparently uncontentious fashion one of the basic tasks of the museum: to show objects to visitors. However, this approach to pedagogy has been severely criticized. The reservations concern the limited view of the processes of communication, and the conceptualization of the receiver/learner. A consideration of these matters in the context of the design and development of museum displays opens up an analysis of the museum-audience relationship, begins to explain why it is in many cases unsatisfactory, and suggests some potential points of change.

The transmission model suggests that the processes of communication and education are one-way, are single, and are linear. The processes are not interactive, and as there is therefore no evaluation or feedback, the responses of the receivers/learners are not considered. The source of the message to be transmitted is the communicator, in the case of the museum, the curator. The curator is the single source of authority and knowledge, and it is his or her task to research the collections, and to put the results of this research, along with illustrative artifacts and specimens, into an exhibition form. It is assumed that a single message will be uniformly received by all visitors, and that the passage of the message, if the exhibition is well designed, is linear and straightforward. Receiver/learners are seen as cognitively passive (empty vessels), and their own knowledge, either everyday common-sense knowledge, or their previous experience, is not recognized as useful or relevant. In the absence of feedback, their response is unknown and unable to be drawn into the communicative or learning process.

Communication, or pedagogy, is limited to the transmission of information only, with other matters screened out. The mass communication model was designed to explain technological processes of communication, such as that of telephonic communications. This understanding of communication does not acknowledge the complexity of human relationships, or begin to consider issues such as gendered or cultural difference. In terms of pedagogy, a separation is made between cognitive and affective aspects of learning processes. Teaching content is completely separate from matters of the everyday. The student's motivation and interest in the information transmitted is assumed. Individual levels of interest or knowledge are unconsidered.

In this approach to museum display, the individualized learning or interpretive processes that might be used by museum visitors in the consumption of the display are not considered. Displays are designed for "the general public"—a mass, passive, abstract, and undifferentiated group.[19] The focus for the curator is, firstly, to develop the best scholarly level of knowledge, on which the display is based; and, secondly, to design the display in such a way that the scholarly knowledge is displayed with conceptual clarity. Pedagogy itself, as a body of practices and forms of expertise, with its own knowledge base that might be applied in the museum, is not considered.

In many traditionally displayed museum galleries, the transmission approach to pedagogy can be observed. Specialist knowledge and curatorial authority inform the design and presentation of the displays that therefore lay out the themes and parameters of specific specialist areas of knowledge with no mediation in terms of communication. For example, at a large city center museum in North America, I recently observed a display of starfish headed *Phylum Echinodermata*. The display is based on the scientific classification of certain organisms, and this is presented as it would be in a formal textbook. The museum speaks to the visitor at a distance, as a gowned university lecturer might ad-

PLATE 1. An example of the transmission approach to display.

dress a group of students in a lecture-hall. This might be compared with new natural history displays at the Auckland Museum in New Zealand, which also display starfish using the taxonomic method, but which also present a reconstructed beach habitat, and living starfish in an aquarium.

Where no evidence to the contrary can be observed,[20] visitors are almost certainly conceived of as the "general public." The facts are presented, and may be absorbed by visitors, if they themselves can make connections with the material provided. The language used is academic, with the aesthetic code and/or the scientific register taken for granted. The collections may be enjoyed if visitors understand their context or appreciate their aesthetic qualities. Whether this is done or not is unknown, as there is no audience testing and therefore no information about the effectiveness of the display. Within exhibitions of this kind, where visitors have not actively been considered, they are positioned as passive receivers. The success of the exhibition pedagogy is uncertain, rather than planned.

Museums which work with a transmission pedagogy are still operating within the modernist paradigm. They have failed to notice, or refused to acknowledge, that the authority of the museum can no longer be taken for granted, and indeed that museum visitors today do not always expect or want to be treated as passive absorbers of factual information.[21] The modernist approach expects its own perspective on history, society, culture and science to be accepted, although within each of these areas specific challenges have been leveled at museum narratives.[22] The transmission approach to pedagogy holds the museum and the audience apart from each other, values the curator as a knowledgeable specialist scholar and the audience as a faceless mass, and leaves the development of exhibitions in the hands of the experts, the museum professionals.

Some museums have accepted that the transmission approach is no longer appropriate and are beginning to develop new relationships with their audiences that position their differentiated publics as collaborators and consultants. The cultural approach to pedagogy embodies these museum/audience relationships, and it is to an examination of this that we now turn.

MUSEUM PEDAGOGY AS PART OF CULTURE

Contemporary ways of understanding communication focus on the meaning-making activities of individuals and groups and see communication as integral to the production and reproduction of culture. Carey describes this second broad approach to communication as a form of secular ritual, as a process of sharing, participation, fellowship and association.[23] Carey's view of communication is overly harmonious, as Giroux points out.[24] Many cultural and communication theorist agree, however, that communication is best understood as a series of processes and symbols whereby reality is produced, maintained, repaired and transformed.[25] Symbolic systems (art, journalism, common sense, mythology, science, museums) shape, express and convey attitudes and interpretations of experience. As beliefs and values are represented through cultural symbols (words, maps, models), so realities are constructed. The realities experienced by different individuals, groups and communities are differentially shaped by the experience of class, gender and race.

The emphasis on communication from this perspective is on its symbolic and interpretive potential. Where the concept of communication as transmission[26] focused on the most efficient way to transfer ideas across space, frequently in the context of the control of people or distance as part of society seen as a political or economic organization,[27] the cultural view of communication addresses the methods by which what counts as "common sense," as "art," or as "science" at any one time are brought into being. Within this view, communication is a much broader process, one which examines ideas in their historical, social and institutional matrices.[28] The significance of communication is as an integral part of culture. Culture itself arises from and is embedded within words, images, symbols, ideas and actions that result in social effects. Naming, classifying and displaying, the basis on which museums operate, have what Hacking has called "looping effects,"[29] tacit or explicit choices made by people to adapt or resist classifications that affect their identities.

The view of communication as culture can be linked to some of the ideas of hermeneutics. Much of the focus of hermeneutics, or interpretive philosophy, concerns how understanding can be achieved. This centers on processes of "interpretation," which is located in the relationship between, typically, a reader and a text.[30] If "museum visitor" and "museum object" (or even "museum") are substituted for "reader" and "text," we can begin to work out a new way of understanding "interpretation" within the museum context—a way which enables a deeper investigation of the processes by which visitors make meaning within museums.

Gadamer's "hermeneutic circle"[31] describes the way in which meaning is made from objects. It is like a conversation. The process of making meaning moves both between the whole and the part of the object and between the present and the past, simultaneously. A dialogue is established between the whole and the part, the past and the present, which enables continual checking and rechecking, revision of ideas, trying out of new ones and rejection of those that don't work. The process of constructing meaning from a painting, for example, is circular and dialogic. We are in a question-and-answer mode, a continuous process as the answers build on those questions that have already been asked and answered. This circular movement involves both the whole and the part, but also the present and the past. Meaning is constructed through this circular action, with modifications to the sense constructed being made constantly.

The content of the conversation is informed by the prior knowledge and experience held by the viewer/interpreter. Depending on the skills and knowledge that interpreters bring to the encounter with an object (an idea, a process), a different conversation will be sustained. In relation to the painting of Van Gogh's *Sunflowers*, for example, an art historian who had studied the period and the artist would have a conversation with the painting that would call on the artist's life-history and his relationship to other artists, and would link this specific painting to others similar to it. A conservator might have a different kind of dialogue with the painting. He or she might ask questions about the light levels to which the painting was exposed, might look at the condition of the canvas, the paint or the frame, might place the painting within a context of the science of paint and the development of yellow pigments. An artist who specialized in painting flowers might interpret *Sunflowers* to discover what he or she could learn about how to paint petals, or flowers in depth, or different tones of yellow. Thus the object might be examined from a number of perspectives, and understood in a range of ways (none of which exhaust the meaning potential of the object) depending on prior knowledge, level of understanding, and intention. In this sense, knowledge is plural, and positioned. Meaning is constructed in the relationship between the "text" and the "reader."

This approach to knowledge and understanding is very close to that of constructivist learning theorists. Both propose that knowledge emerges through the interpretation of experience made by the knower. Both assert that knowers, or learners, or interpreters, are active in the processes of making sense of experience (including the formal or informal experience of learning). Both mental and bodily actions are important in learning processes, and learners will use differentiated learning strategies to both perceive and process information and experience. Individual learners will combine concrete and abstract modes of perception, and active and reflective processes of accommodating and assimilating knowledge, according to their preferred approach to learning.[32]

As we have seen, both hermeneutics and constructivist learning theory assert that knowers, or learners, are active in the processes of making sense of experience. However, meaning-making is not an entirely individual process. Although each of us is active in interpreting our own experience, using our individual strategies, capabilities, and preferred learning styles, the interpretation that we make is not ours alone. It is mediated, tested and developed within the context of communities of interpretation. Constructivist learning theory tells us that learning is both personal and social.[33] The concept of "interpretive communities" enables a focus on how this social aspect to learning might be understood.

INTERPRETIVE COMMUNITIES

The meanings constructed from information and experience emerge from a complex network of mediation. Personal interpretations are forged through social and cultural environments, through local communities and through location in social structures. Although none of these elements are immutable, personal meanings and interpretations do have social dimensions. Individual meaning-making is tested in the context of interpretive communities. What we "know" is produced through interpretations of individual experience, but also through the testing and refining of our interpretation within those communities that are significant to us.

Stanley Fish,[34] a literary theorist, defines an "interpretive community" as all those individuals who jointly share the same strategies for reading texts and assigning meaning.[35] He suggests that these strategies exist prior to the act of reading and therefore determine the shape of what is read. The interpreting reader, acting with purposes and concerns, which have emerged through pre-existing experience and learning, determines what counts as the facts to be viewed as significant in a text.[36] Thus with Van Gogh's *Sunflowers*, art historians, conservators and flower-painters would consider different facts to be salient. Different categories of understanding would be employed because of local relevance to the interpreter.

An interpretive community can be identified where the same systems of intelligibility are being used, where the same categories of understanding are employed and the same significant entities are recognized. In other words, those who share cognitive frameworks, attitudes and values, and interpretive or analytical skills, would very likely belong to the same interpretive community. Art historians, conservators and flower-painters would belong to different interpretive communities. Fish wishes to show how meaning is not inherent in the text (or we might substitute—the object or exhibition), but is brought into being through the interpretive strategies used by the reader (viewer, visitor). These meanings are in large part controlled by the strategies sanctioned and the validation accorded them by the relevant interpretive community. It is the authority of the interpretive community that enables meaning. At the same time, interpretations differ between interpretive communities. Different systems of knowledge (art, science, history), different ontological or epistemological perspectives in relation to these systems, cultural or ethnic differences, and gendered approaches, create differentiated communities whose interpretations of their experiences vary. Struggles over the validity of different explanations are part of the character of society, and are carried out in the context of power relationships that are not, in themselves, limited to the cultural.

Although the concept of interpretive communities is within some fields (media studies, English) a familiar one,[37] and it is a slippery concept to apply within empirical research studies,[38] the outline concept is useful in a discussion of museum communication and pedagogy. It highlights the significance of differentiated meaning-making strategies and repertoires. Fish, being concerned with the way in which meaning is constructed from literary texts, is concerned with the different readings that different specialist knowledge might produce. This has particular resonances within the museum field.

In the museum, the issue of specialist knowledge is important. If curators as members of art or science communities interpret the museum collections from their own perspectives, they are likely to communicate only within their own interpretive communities. To make the interpretation relevant outside of this community it is necessary to discover what interpretive strategies are employed, what levels of specialist knowledge have been attained, and indeed whether there is any knowledge at all of specific art or science concepts. Clearly this will vary among different groups.

Research shows that where opportunities are offered for the deployment of a range of levels of knowledge and a range of preferred learning styles, much greater learning occurs among more people. Modifications to the New England Lifezones Hall at the Boston Museum of Science included the provision of more accessible labels, mounted specimens or sculptures of animals that could be touched, smell-boxes and small items that could be manipulated. After the modifications, the time spent by visitors in the gallery increased, the gallery was used in a more diverse fashion, the use of varying learning styles were observed, and more factual information was recalled.[39]

At the new National Museum of New Zealand Te Papa Tongarewa, the language used in the display texts is vibrant, contemporary and imaginative. It

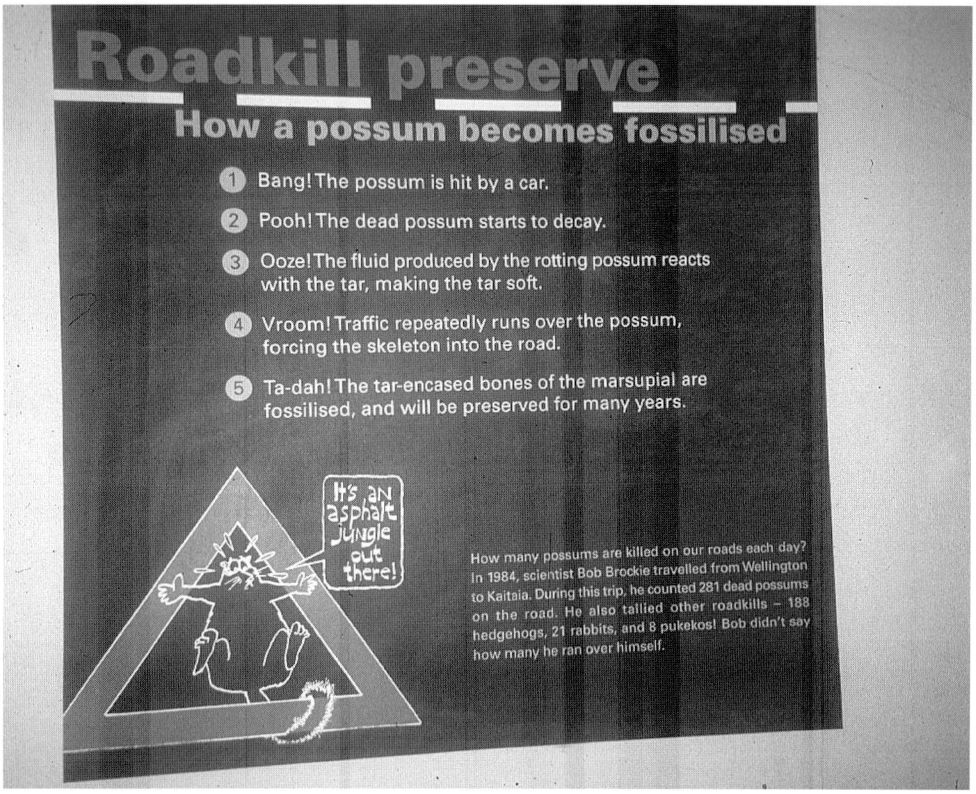

PLATE 2. An example of the cultural approach to display.

enables the museum to speak directly to its visitors as though to a friend or neighbor in an informal everyday environment.

In the second approach to communication, the focus is on the meanings constructed from experience by the individual, within a social network. Meanings are negotiated through cognitive frameworks, interpretive strategies and interpretive communities. Meanings are accepted as plural and open to challenge; it is acknowledged that audiences are active,[40] and that learners find their own ways of knowing and explaining the world. Within this second approach to communication and learning, teacher/communicators act as enablers and facilitators. The task for teachers, communicators, or museum exhibition developers is to provide experiences that invite visitors to make meaning through deploying existing interpretive strategies and repertoires. Their role is to provide environments for effective learning, where learners can explore according to their prior knowledge and their preferred learning styles, and where they can test their hypotheses against those of others, including those of experts in the field. The task is to present opportunities for visitors to use what they know already to build new knowledge and new confidence in themselves as learners.

To understand pedagogy as part of the construction and reconstruction of culture is to open the museum to the multifaceted struggles for acceptance and validity of the everyday. It means assessing the museum from multiple differentiated perspectives that operate from different levels of social power. And it means an awareness of the necessary enmeshing of the museum within the challenges of cultural politics. Nothing else is possible any longer.

NOTES

1. Chris Smith, "Arts for the many, not the few, Who goes to the arts?" *The Independent*, special report (March 30, 1999), pp. 2–3.
2. *A New Cultural Framework* (London: Department for Culture, Media and Sport, 1998).
3. Ibid.
4. Victor Middleton, *New Visions for Independent Museums in the UK* (West Sussex: Association of Independent

Museums, 1990); Victor Middleton, *New Visions for Museums in the 21st Century* (London: Association of Independent Museums, 1998).

5. See Eilean Hooper-Greenhill, *Museums and the Shaping of Knowledge* (London: Routledge, 1992) for a discussion of the changing concept of a museum since the Renaissance.
6. Philip Hensher, "I was wrong—we should hand back the Elgin Marbles," *The Independent* (August 27, 1999), p. 4.
7. Andreas Huyssen, "Escape from Amnesia: the Museum as Mass Medium," in *Twilight Memories: Marking Time in a Culture of Amnesia*, ed. Andreas Huyssen (New York and London: Routledge, 1995), pp. 1–35.
8. Johan Fornas, *Cultural Theory and Late Modernity* (London and New Delhi: Sage, 1995), p. 19.
9. Gerard Delanty, *Social Science: Beyond Constructivism and Realism* (Buckingham: Open University Press, 1997), pp. 25–30.
10. Tony Bennett, *The Birth of the Museum* (London: Routledge, 1995).
11. Quoted in Janet Minihan, *The Nationalisation of Culture—the Development of State Subsidies to the Arts in Great Britain* (London: Hamish Hamilton, 1977), p. 26.
12. John Peponis and Jenny Hedin, "The Layout of Theories in the Natural History Museum," *9H*, no. 3 (1982): 24.
13. Philip Fisher, *Making and Effacing Art: Modern American Art in a Culture of Museums* (Oxford and New York: Oxford University Press, 1991), pp. 8–9.
14. Barbara Kirshenblatt-Gimblett, *Destination Culture: Tourism, Museums, and Heritage* (Los Angeles and London: University of California Press, 1988), p. 33.
15. Eilean Hooper-Greenhill, "The Museum in the Disciplinary Society," in *Museum Studies in Material Culture*, ed. Susan Pearce (Leicester: Leicester University Press, 1989), pp. 61–72.
16. Robert L. Heath and Jennings Bryant, *Human Communication Theory and Research: Concepts, Contexts and Challenges* (Hillsdale, N.J., Hove and London: Lawrence Erlbaum Associates, 1992); Denis McQuail, *Mass Communication Theory—an Introduction*, Second Edition (London: Sage Publications, 1987).
17. Denis McQuail and Sven Windahl, *Communication Models for the Study of Mass Communication*, Second Edition (London and New York: Longman, 1993), pp. 8–61.
18. Eric Sotto, *When Teaching Becomes Learning: a Theory and Practice of Teaching* (London and New York: Cassell, 1994).
19. Eilean Hooper-Greenhill, "Learning from Learning Theory in Museums," *GEM News*, no. 55 (Group for Education in Museums, 1994), pp. 7–11.
20. Such as provision for specific physical needs or for differentiated learning styles; for example, thermoforms (raised pictures) for people who are visually impaired, chairs for the elderly, or touch-carts for those who prefer to learn enactively, such as children or those new to the subject matter).
21. See, for example, Yvonne Harris and Ben Gammon, *Join the Great Fat Debate*, unpublished internal Science Museum report, 1998. This research showed that "visitors were happy for the Museum to cover topics to which there is no definite answer. Some visitors expected the Museum to do this. Visitors were happy to make up their own minds as to whether or not olestra is safe to eat, rather than being provided with a definite answer by the Museum. Most visitors were not surprised that the Museum has asked visitors for their opinion as part of the exhibition." Details supplied by Xerzes Mazda and Ben Gammon, August 1999.
22. See, for example, some of the papers in the following anthologies: *Contesting Art: Art Politics and Identity in the Modern World*, ed. Jeremy MacClancy (Oxford and New York: Berg, 1997); *Representing the Nation: a Reader—Histories, Heritage and Museums*, eds. David Boswell and Jessica Evans (London and New York: Routledge, 1999); *Theorizing Museums*, eds. Sharon MacDonald and Gordon Fyfe (Oxford: Blackwell, 1996); *The Politics of Display: Museums, Science, Culture*, ed. Sharon MacDonald (London and New York: Routledge, 1998).
23. James Carey, *Communication as Culture: Essays on Media and Society* (Boston: Unwin Hyman, 1989), pp. 13–36.
24. Henry Giroux, *Border Crossings: Cultural Workers and the Politics of Education* (New York and London: Routledge, 1992), p. 168.
25. Carey, p. 23.
26. Carey, p. 22, points out that the dominant understanding of communication in American public life has been as transmission. Cultural theory as a whole is less well developed than in Britain and Australia. See Graeme Turner, *British Cultural Studies: an Introduction* (Boston: Unwin Hyman, 1990).
27. Carey, p. 15 and p. 34.
28. Ian Hacking, *The Social Construction of What?* (Cambridge, MA and London: Harvard University Press, 1999), pp. 10–11.
29. Ibid., p. 34.
30. Shaun Gallagher, *Hermeneutics and Education* (New York: State University of New York Press, 1992).

31. Hans-Gorg Gadamer, "The Historicity of Understanding," in *Critical Sociology—Selected Readings*, ed. Paul Connerton (Harmondsworth, Mx: Penguin Books, 1976), pp. 117–133; Eilean Hooper-Greenhill, "Education, Communication and Interpretation: Towards a Critical Pedagogy in Museums," in *The Educational Role of the Museum*, Second Edition, ed. Eilean Hooper-Greenhill (London: Routledge, 1999), pp. 3–27; Eilean Hooper-Greenhill, "Learning in Art Museums: Strategies of Interpretation," in *The Educational Role of the Museum*, Second Edition, ed. Eilean Hooper-Greenhill (London: Routledge, 1999), pp. 44–52.
32. Bernice McCarthy and Susan Leflar, *4mat in action: creative lesson plans for teaching to learning styles with right/left mode techniques* (Barrington, Illinois: Excel, 1983); Gunther F. Charles, "Museum-goers: lifestyles and learning characteristics," *The Educational Role of the Museum*, First Edition, ed. Eilean Hooper-Greenhill (London: Routledge, 1994), pp. 286–297.
33. Falk and Dierking explain this in relation to learning in the museum. See John Falk and Lynn Dierking, *The Museum Experience* (Washington: Whalesback Books, 1992), pp. 1–7. See also George Hein, *Learning in the Museum* (London: Routledge, 1998) for a comprehensive discussion of constructivist and realist educational theories in the museum context.
34. Stanley Fish, *Is there a text in this class? The authority of interpretive communities* (Cambridge, MA and London: Harvard University Press, 1980).
35. Ibid., p. 10.
36. Ibid., p. 8.
37. Kim Christian Schroder, "Audience semiotics, interpretive communities and the 'ethnographic turn' in media research," *Media, Culture and Society*, vol. 16 (London, Thousand Oaks and New Delhi: Sage, 1994): 337–347.
38. Jonathon Potter and Margaret Wetherall, *Discourse and Social Psychology: Beyond Attitudes and Behaviour* (London: Sage, 1987), pp. 138–157.
39. Betty Davidson, Candace Lee Heald and George Hein, "Increased Exhibit Accessibility through Multisensory Interaction," in *The Educational Role of the Museum*, Second Edition, ed. Eilean Hooper-Greenhill (London: Routledge, 1999), pp. 223–241.
40. The "active audience" is a long-standing concept within media studies. See David Morley, "Theories of consumption in media studies," in *Acknowledging Consumption: a Review of New Studies*, ed. Daniel Millar (London and New York: Routledge, 1995), p. 310.

Concluding Remarks

John L. Heilbron

Exhibiting and exhibitionism are high-stake games, capable of exercising great influence on public opinion and dependent increasingly on the approbation of public officials. The games are complicated by the diversity of services, approaches, and purposes characterizing science museums and science centers; by disparity in their sizes and resources; by the uncertainty and opportunity introduced by computers; by the tension between maximizing the quantity of visitors and the quality of their visit; and, no doubt, by a thousand other considerations. When I add all these together, I calculate that modern museology must be a fertile field for contention. I was therefore extremely surprised that the authors of the articles seem, at first, to agree about almost everything.

While considering this oddity, I recalled Eric Hobsbawm's depiction of his colleagues as "peaceful herds of historians grazing in the rich pastures of their primary sources and chewing the cud of one another's publications." But a second reading showed that this is not the case. My summary is an attempt to draw out some of the strong oppositions that I think I have found. I've arranged my remarks under the neutral headings of coercion, guile, and arrogance.

COERCION

Jim Bennett gives an eloquent plea against reducing the message of exhibitions to currently accepted scientific dogma. Instead, the exhibition of an instrument or some other relic of past science should stimulate a variety of responses. The object can be appreciated as a cultural artifact, a material thing, an embodiment of social relationships, and so on, as well as the incorporation or instantiation of some scientific principle or mathematical representation. But he does not tell us how much leeway he wants to grant his visitors. Simon Schaffer warns that a museum's assemblage of objects can inculcate comprehensive world views; "object lesson and social order are produced together," he says; visitors receive simultaneously instructions about "the correct order of nature and of society."

Let us accept Schaffer's bold claim. Where does it leave Bennett? If he wants to inhibit his visitors from devising world views he thinks wrong and even immoral, if there are any such, he must either disagree with Schaffer or try to limit the range of interpretations to which his displays might easily give rise. This is not an idle observation. A museum dedicated to the Holocaust would not arrange its material so as to provide the slightest support for the neo-Nazi claim that there was no Holocaust. Nor—to bring the question closer to home—would a science center consciously propagate the view that, on balance, science is inimical to society.

Ben Gammon and Xerxes Mazda provide us with an arresting breadth of laissez-faire. Visitors to their exhibits on such sensitive issues as the use of genetically modified foodstuffs are invited to express their opinions, usually ill-informed, in writing; and, provided these opinions are not blasphemous, irrelevant, or smutty, they are reproduced for other visitors to read. Most of the written comments oppose introducing genetically modified materials into the food chain. If the Science Museum plays this game honestly, it has very little control over the message it wants to impart—if it has a message at all. And, as Gammon and Mazda observe, one can imagine that the opinions thus collected might achieve a wide and baleful influence if reported in the press.

In contrast to this permissiveness, Alan Friedman tells us that as he has put more structure into visits to his museum, visitors learn more of the material he wants to teach them. Guided audio tours have proved the best way to keep minds concentrated for more than the minute or two that is the average stay before a display when people wander through his museum on their own. Friedman suggests that, as far as he has gone, the more the control the better the result in a science center. Is this finding peculiar to his sort of museum and inapplicable to Bennett's sort? What is the desirable degree of latitude? How can it be secured?

GUILE

Ken Arnold guilelessly writes that he uses guile to attract visitors to his exhibitions of science and art; he also said that he "exploits images to arrest attention." Neil Cossons talked about shopping-mall technologies and locations to lure in idle shoppers. Wolf Peter Fehlhammer arouses emotion in his visitors by scattering works of art among his artifacts. Dominique Ferriot and Bruno Jacomy approved Fehlhammer's the plan and say that objects should be selected for display not primarily for their rarity or scientific interest, but for the emotional response they elicit.

Perhaps Ferriot and Jacomy had in mind the building housing their collections, which in itself excites the imagination. An edifice built for the needs and purposes of religion, converted by Revolution to serve the cult of machines, now shelters the relics of obsolescent modernism. Ferriot's impressive reworking of this ancient space is full of appeal to the several relevant parts of the complete visitor—to the mind and the eye, as well as to emotion and imagination.

Another sort of emotional appeal is practiced in Catalonia. Jorge Wagensberg explicitly targets the emotion of intelligibility, the recognition of sameness in difference, the charm of perceiving the familiar in what at first seems strange and extravagant. In the Science City in Calcutta, as Saroj Ghose made clear, inter-active exhibits engage the flighty visitor. There are many ways to the *captatio benevolentiae* needed to move traffic into the museum.

Although it is desirable to exhibit objects that immediately claim the attention of visitors, that does not mean—as I am sure none of the authors mean—that anything goes. Yet they go far enough. One example—a little unfair, perhaps—will allow me to make my meaning clear. I have in mind and in my mind's eye the torsos rising from atomic bombs and the carpets figuring the destruction of Dresden deployed in the airplane wing of the Deutches Museum. An unfriendly reading of this juxtaposition is that Germany and Japan were the victims of World War II. I make this observation not because I believe for a moment that this was the message intended, but to enforce the view that it is even harder to control responses to works of art than to objects of science.

These measures—the appeal to the emotions and the use of guile—seem to be motivated by the felt need to keep up the number, and enlarge the spectrum, of visitors. Perhaps the importance of grasping for visitors has been exaggerated. Neil Cossons, whose Science Museum in London is in the van of innovative methods of population increase, writes that he has had to agree to meet certain targets in exchange for state aid. The context suggests that the targets are visitor numbers. In fact, the human throughput is but one of a dozen performance objectives, another of which is the amount of peer-reviewed published research; no doubt performance in the several domains can be played against one another and quality can raise its head against quantity in the final reckoning. This suggests that adroitly written agreements need not bind museum directors to multiplying gimmicks to multiply attendance. I know from my own experience that administrators often talk themselves out of proposing innovations on the assumption that the official who gives the money will reject them. Ask before surrendering.

Why are there so few sitting places in an exhibition space? Because the opportunity to sit greatly increases the time visitors spend in the museum and also increases the amount they read and take away. Therefore, of course, there are few or no seats at all in science museums. People who dally may learn more, but they decrease the throughput. Hence the reductio ad absurdum of seductio ad nauseum: visitors are lured into the museum with great ingenuity in order to be ejected with great alacrity. By this measure, the greatest transfer of funds comes with the least transfer of information.

ARROGANCE

As Bernard Finn points out, the tacit assumption that the curator knows best, which underlays several articles, has a strong air of arrogance. In this connection, as cautionary tales, I will enter briefly into that cornucopia of lessons in museology, the Smithsonian Institution of Washington. I have in mind the contentious exhibit, "Science in American Life," and the contentious non-exhibit, "The Flight of the *Enola Gay*." Both are mentioned by Fehlhammer and Finn with the intimation that they involved violations of academic freedom.

"Science in American Life" was paid for by the American Chemical Society, which agreed that the Smithsonian would have the right to decide the exhibit's content. Although representatives of the society sat on the committee that advised the curators, they did not succeed in obtaining a script satisfactory from their point of view. True to their word, they paid their bill and did not protest, although in their view the exhibit was unfairly critical of science. The American Physical Society was not similarly muzzled. Its officers, including a winner of the Nobel prize, objected to the Smithsonian's Secretary that "Science in American Life" failed to convey their feeling of the excitement of discovery and most people's appreciation of the improvement of life brought about by science-based technology. Instead, in their opinion, the exhibit dwelt unjustifiably on industrial pollution, the evils of pesticides, and the horrors of the atomic bomb.

The Secretary assembled a committee on which the physicists, a few Smithsonian administrators, and several consultants, including Alan Friedman and myself, sat. The committee diagnosed a few ills and recommended cures; the Smithsonian agreed to them; the storm dissipated; and, as far as I know, nothing of significance changed. The problem with the exhibit, in my opinion, was not that it offended the scientists, but that it was (and is) poor history. As the survival of "Science in American Life" virtually intact and the subsequent advancement of its head curator show, poor history at the Smithsonian is not necessarily fatal to its practitioners. But it can be. That brings us to the *Enola Gay*.

The *Enola Gay* was the plane that carried the atomic bomb to Hiroshima. Neglected after the war, the bomber was eventually refurbished by the Smithsonian under pressure from veterans' groups. The Smithsonian's Air and Space Museum proposed to exhibit the *Enola Gay*'s fuselage in commemoration of the fiftieth anniversary of the end of World War II.

The first draft of the script for the exhibit, the work of revisionist historians, made Japan appear to have been the primary victim of the second world war. The Director of Air and Space, realizing that the script was imperfect, gave copies to veterans' groups to review, anticipating, in his naiveté, a scholarly exchange of views. The veterans did not join the peaceful grazing in the historians' pastures. They went to Congress rather than the library and shot down the *Enola Gay*. Although the script was rewritten to give Japan due credit for World War II, the veterans were unappeasable; the Secretary of the Smithsonian believed that he would endanger his institution by insisting; he canceled the exhibit, and six months later, fired the director of Air and Space.

In both "Science in American Life" and "The Flight of the *Enola Gay*," historians with views similar to those of the museum's curators advised about content without taking sufficiently into account the views of other interested parties. They made the error of thinking that historians own recent history as we do most of the past. Usually we can graze our pastures and chew our cud without external molestation; that is, regrettably, because usually no one else cares what we do. Matters change when the historical actors, or their widows and children and pious students, are still alive. It is arrogant as well as foolhardy to disregard their understanding of their experiences. We may judge their recollections to be compromised by false consciousness, selective memory, or interested interpretation; but often we do not know for certain that they are wrong and, in any case, their recollections and convictions are part of the record we must interpret. We should accommodate the retrospective understanding of former actors in our exhibitions not as a *pis aller*, or as a defense, but because it has a place there and may teach us something about the subject under study. Modern history is still a battlefield as, indeed, it should be: our conception of our recent past, true or false, determines our attitudes toward contemporary problems just as our experiences in our society affect our historiography.

The Science Museum deserves applause for its new close cooperation with scientists at their principal outlet, the journal *Nature*. The Deutsches Museum appears to have a similar program in view. In addition to scientists, science museums should coopt in the design of major exhibitions other principal constituencies for whom the history to be reconstructed still has life. Eilean Hooper-Greenhill ends her article with much the same consideration. I hope that her advice will be heeded.

Chapter opening illustration, *The Thinker* (detail) by Auguse Rodin, 1880. *Courtesy The Rodin Museum, Philadelphia: Gift of Jules E. Mastbaum.*

Photo: Graydon Wood 1996

Notes on Contributors

Ken Arnold studied natural sciences at Cambridge and wrote his Ph.D. dissertation on the history of English museums for Princeton University. Since then, he has worked in a variety of museums on both sides of the Atlantic. His arrival at the Wellcome Trust in 1992 allowed him further to pursue his historical interests, while enabling him to present his findings in the form of exhibitions. More recently he has taken up the position of Exhibitions Manager at the Wellcome Trust, overseeing two galleries dedicated to exhibitions dealing with the culture of medicine (its science, art and history) and project managing a variety of initiatives that bring together science and art, most notably the funding initiative sciart. He also writes and lectures on the culture of museums past and present.

Jim Bennett is Keeper of the Museum of the History of Science in Oxford and is President of the British Society for the History of Science and of the Scientific Instrument Commission of the International Union of the History and Philosophy of Science. He works on the histories of scientific instruments, of astronomy and of practical mathematics, as well as on the uses of museums.

Sir **Neil Cossons** has been Director of the Science Museum, London—The National Museum of Science & Industry—since 1986. After his retirement in 2000, following a museum career which began in 1961, he will become Chairman of English Heritage, the government body responsible for the protection and preservation of historic buildings and ancient monuments. Sir Neil has written and broadcast widely on museums, industrial archaeology and the public understanding of science.

Wolf Peter Fehlhammer has been General Director of the Deutsches Museum in Munich since 1993, and President of ECSITE, the European Collaborative for Science, Industry and Technology Exhibitions, since 1998. He is also Honorary Professor of Inorganic Chemistry at the University of Munich. Between 1983 and 1993 he held a chair in Organometallic Chemistry at the Free University of Berlin, and between 1976 and 1983 an associate professorship at the University of Erlangen-Nürnberg.

Dominique Ferriot—a university professor—has been Director of the Musée des Arts et Métiers in Paris since 1988. She initiated and is currently directing the renovation project underway as part of the French State's "Grands Travaux" program of public works. Her previous posts have included Assistant Director of the Écomusée in Le Creusot (1976–1979), Head of partnership operations and regional action in the team developing the future Musée de la Villette (1980–1984), Director of Scientific and Technical Culture at the Ministry of Research and Technology (1985-1987). Her work was originally focused on industrial archeology and later became centered on technical museology: organization of colloquiums and creation of *La Revue*, a quarterly publication devoted to the history and museology of technologies.

Bernard Finn is Curator of the Electrical Collections at the Smithsonian Institution's National Museum of American History. His research interests have focused on 19th-century electrical technology and, in particular, on submarine telegraphy. Having written on the history of technical museums, he has also been responsible for some two dozen exhibitions, most in allied subject areas. Currently he teaches a seminar on museums as part of a joint program with the University of Maryland, and is co-editor of the annual publication *Artefacts*, jointly sponsored by the Science Museum, the Deutsches Museum, and the Smithsonian.

Robert Friedel is Professor of History at the University of Maryland, where he teaches the history of technology and science. Most of his research and writing have been related to the history of invention and materials, with books that include *Pioneer Plastic: The Making and Selling of Celluloid* (Wisconsin, 1983) and *Zipper: An Exploration in Novelty* (W. W. Norton, 1994). He has been a historian at the Smithsonian Institution's National Museum of History and Technology and Director of the IEEE Center for the History of Electrical Engineering, and has worked on exhibitions at the National Museum of American History and the National Building Museum, Washington.

Alan J. Friedman has been the Director of the New York Hall of Science, New York City's public science-technology center, since 1984. Before coming to New York he served as Conseiller Scientifique et Musèologique for the Cité des Sciences et de l'Industrie, Paris, and was the Director of Astronomy and Physics at the Lawrence Hall of Science, University of California, Berkeley for 12 years. He is the co-author, with Carol C. Donley, of *Einstein as Myth and Muse* (Cambridge University Press, 1985). The American Association for the Advancement of Science named him the 1996 recipient of its Award for Public Understanding of Science and Technology. He received his Ph.D. in Physics from Florida State University and his B.S. in Physics from the Georgia Institute of Technology. He is a Fellow of the American Association for the Advancement of Science and of the New York Academy of Science.

Paolo Galluzzi is Professor of the History of Science at the University of Florence and Director of the Istituto e Museo di Storia della Scienza in Florence.

Ben Gammon is Head of Visitor Research at the Science Museum, London. He has worked on many exhibition teams as an audience advocate promoting a visitor-centered approach to interpreting science and technology. He is currently working on the Wellcome Wing project —a new suite of exhibitions about contemporary science and technology which opens in summer 2000.

Saroj Ghose, the immediate past President of ICOM, the International Council of Museums in UNESCO 1992-98, and the retired Director General of the National Council of Science Museums of India 1979-97, has been instrumental in developing a large chain of 26 interactive science centres all over India including the Science City at Calcutta. At present he is working as Adviser to develop another giant science city in Ahmedabad and high-tech story-telling museums at Calcutta and Punjab. For his contribution to science popularization, he has received awards from the President of India, the Indian National Science Academy, University Grants Commission, and Trieste International Foundation in Italy. He has been awarded a D.Sc. (Honoris Causa) from North Bengal University.

Marika Hedin has a B.A. from Stockholm University and an M.A. from the University of Notre Dame. She is a Ph.D. candidate in History at Stockholm University. Since 1998 she has been a Curator at the Nobel Museum.

John L. Heilbron, formerly Professor of History and Vice Chancellor at the University of California, Berkeley, is Senior Research Fellow at Worcester College, Oxford.

Eilean Hooper-Greenhill is Professor and Director of the Department of Museum Studies at the University of Leicester. Her research interests focus on the relationship of museums to their audiences. In 1999, she opened the Research Centre for Museum and Gallery Education (RCMGE). She is the author of *Museum and Gallery Education* (Leicester University Press, 1991), *Museums and the Shaping of Knowledge* (Routledge, 1992), *Museums and Their Visitors* (Routledge, 1994) and several edited volumes. She is currently completing *Museums, Pedagogy and Social Change: the Politics of Visual Culture* to be published by Routledge.

Bruno Jacomy, an engineer and graduate of the École Nationale Supérieure des Arts et Métiers, started his career as a librarian at the Écomusée in Le Creusot, where he worked from 1976 to 1983. He completed a thesis in social psychology on riveting techniques. From 1984 to 1988, he was responsible for coordinating the "Living Encyclopedia" operation on a national level (exhibitions on technologies dating from Diderot's time to the present day). He published *Une Histoire des Techniques* (Éditions du Seuil, 1990) and, since 1990, has been in charge of the "Technical Culture" course at the University of Technology in Compiègne. Since 1989, he has been Assistant Director of the Musé des Arts et Métiers, and has been coordinating the steering committee for the Museum renovation project.

Ulf Larsson has a Ph.D. in History of Technology from the Royal Institute of Technology in Stockholm. He was a Scientific Secretary at the Swedish Foundation for Strategic Research in 1997–1998. Since 1998, he has been a Senior Curator at the Nobel Museum.

Svante Lindqvist has a Ph.D. in History of Science and Ideas from Uppsala University. He was Professor of History of Technology at the Royal Institute of Technology in Stockholm 1989–1997. In the fall of 1992 he was Visiting Professor at the University of Pennsylvania and in 1995–96 Overseas Fellow at Churchill College, Cambridge. He was elected a member of the Royal Swedish Academy of Engineering Sciences in 1992 and of the Royal Swedish Academy of Sciences in 1994. Since 1998 he is Director of the Nobel Museum.

Xerxes Mazda is a Curator and Exhibit Developer at the Science Museum, London. He has principally worked on exhibitions about chemistry, materials and the environment. He is currently working on an ambitious series of discussion exhibits for the Wellcome Wing project.

Simon Schaffer is Reader in History and Philosophy of Science and Fellow of Darwin College, University of Cambridge. He has helped organize exhibitions in Cambridge and London on nineteenth-century physics and its public image ("Empires of Physics,"1994), on science at the Paris 1900 exhibition ("The New Age," 1995), and on digital technology and art ("N01SE,"2000). His most recent publication is the jointly edited collection, *Sciences in Enlightened Europe* (Chicago University Press, 1999).

H. Otto Sibum is Research Director at the Max Planck Institute for the History of Science, Berlin and Affiliated Research Scholar of the Department of History and Philosophy of Science at the University of Cambridge. He is studying the history of experimentation in the physical sciences from the eighteenth to the twentieth century.

Doron Swade is Assistant Director and Head of Collections at the Science Museum, London. He is an engineer and an historian of computing, and has published widely on the history of technology and curatorship. His most recent book (Spring 2000) is *The Cogwheel Brain: Charles Babbage and the Quest to Build the First Computer.*

Jorge Wagensberg was born in Barcelona in 1948. In 1971 he graduated in Physics and in 1976 obtained his doctorate at the University of Barcelona, where he now lectures in the Theory of Irreversible Processes at the Faculty of Physics and directs a biophysics research group. He is the author of many scientific articles published in international specialist journals and of an extensive body of work aimed at popularizing science in other areas of culture. In 1980 he published the book *Nosotros y la ciencia* (Bosch Editor) and in 1985 *Ideas sobre la complejidad del mundo* (Tusquets Editores) — translated into French: *L'âme de la méduse* (Le Seuil, 1997). His other books include *Introdocció a la Teoría de la Probabilitat i la Informació* (Edicions Proa, 1996), *Amazonia, el último paraíso* (Ambit Serveis Editorials, 1996), *Ideas para la Imaginación Impura 53 reflexiones en su propia sustancia* (Tusquets Editores, 1998). In 1983 he created the "Metatemas" series on scientific thought (Tusquets Editores). Since 1991 he has been the Director of the Museo de Ciencia de la Fundación "La Caixa."

Index of Names